STRATA PUBLISHING

AN INTRODUCTION TO SMALL UAS DEPLOYMENT FOR EMERGENCY RESPONDERS

Considerations for Organizational
Requirements and Operations

Gus Calderon • Kevin Rolfe • Jonathan Rupprecht

Copyright © 2017
Authors: Gus Calderon, Kevin Rolfe, Jonathan Rupprecht
Editor: Maha Calderon

All rights reserved. No part of this book may be reproduced in any manner without written permission except in the case of brief quotations included in critical articles and reviews.

Published by Strata Publishing
Printed and bound in the United States
Columbia, Maryland

ISBN-13: 978-1544678924
ISBN-10: 1544678924

1. Technology 2. Aeronautics 3. Government 4. Emergency Management

First Edition First Printing

10 9 8 7 6 5 4 3 2 1

Table of Contents

Introduction 7

Preface 8

Section 1: Introduction to Small UAS 10

    Small UAS Definition 11
    Fixed-Wing 11
    Rotary-Wing 12
    Multi-Rotor 13
    Transition 13
    Airframes & Components 14
    Lithium Battery 14
    Electronic Speed Controller 15
    Motor 15
    Flight Controller 15
    Servos and Actuators 16
    Ground Control Station 17
    Imaging Sensor 17

Section 2: Federal Aviation Administration 18

    Background 19
    Federal Aviation Regulations 20
    FAA Modernization and Reform ACT of 2012 20
    Part 107 Regulations 21
    FAA Enforcement 22

Section 3: National Airspace System 23

    Aeronautical Charts 24
    National Airspace System (NAS) 25
    Controlled Airspace 25
    Uncontrolled Airspace 26
    Special Use Airspace 26
    Other Airspace Areas 28

Section 4: Types of Small UAS Operations 30
Public Operations (Government) 30
Civil Aircraft Operations 31
Model Aircraft (Hobby or Recreational) 32

Section 5: Small UAS for Emergency Management 34
Small UAS Integration 35
Common Operational Picture 36
Small UAS Sensor Technology 36
Small UAS Emergency Management Applications 38
Cost Benefits 41
Community Engagement and Safety of the Public 42

Section 6: Small UAS Program Structure 43
Small UAS Program Structure 44
Program Manager (PM) 44
Operations Manager (OM) 45
Remote Pilot in Command (RPIC) 47
Payload Operator (PO) 48
Visual Observer (VO) 49
Maintenance Technician (MT) 49
Storage and Accountability 50
Small UAS Personal Protective Equipment (PPE) 50
Small UAS Reporting and Data Points 52

Section 7: Small UAS Flight Operations 53
Flight Operations and Procedures Manual 54
Mission Planning 54
Flight Rules 55
Prohibited Operations 55
Preflight Inspection 56
Operational Assessment 57
Briefings and Security 57
Flight Termination 58

Section 8: Small UAS Post Flight Operations 59
Post Flight Mission Report 59
Accident and Incident Documentation 60
Flight Logbook 60
Small UAS Status Tracking Board 61

Small UAS Budget Document 61

Section 9: Small UAS Training **62**
NIMS/ICS Training 63
RPIC Training 64
VO Requirements 66
Payload Operator (PO) Requirements 66

Section 10: Small UAS Maintenance **67**
Scheduled Maintenance 68
Maintenance Personnel 68
Inspection Requirements 69
Record-keeping Requirements 70

Conclusion **71**

References **72**

About the Authors **74**

Introduction

The purpose of this book is to provide guidance to **government entities in the acquisition, integration and operation of small unmanned aircraft systems** (UAS) during emergency response scenarios.

The ability to collect data with remotely controlled ground and marine vehicles has saved organizations considerable time, effort and expense. In August 2016, new regulations were enacted by the FAA to facilitate the civilian operation of small UAS. Hazard mitigation and incident response with small UAS is becoming a more streamlined process.

When deployed by properly trained personnel, small UAS can rapidly provide real-time information at a fraction of the cost. Small UAS are meant to complement the use of conventional aircraft, augmenting data collection capability. Small UAS offer rapid deployment, increased maneuverability, better accessibility and dependable resilience.

Properly trained personnel and a thoroughly maintained small UAS are essential to the implementation of a successful small UAS program.

An organization's specific mission planning will be based on many factors, including the type of small UAS platform and data collection, the level of pilot and flight crew capabilities and experience to the complexity of the mission.

An assessment of the operational and financial risks against the desired benefits of a small UAS program must be based on an understanding of the existing FAA rules and regulations.

The adaptation of small UAS technology can potentially transform the way an organization meets its mission requirements. An organization can achieve this goal either by developing its own small UAS program or by subcontracting the services of a reputable small UAS provider.

Preface

The success of an organization depends on its ability to properly acquire data and assess available resources while remaining in control of operational costs. The need for collecting, analyzing and acting on data has become critical in the information age. Organizations that provide products and services on a day-to-day basis may become affected or impacted by an emergency. Any emergency occurrence may further challenge the organization's ability to manage its costs. The ability to obtain information and develop actionable intelligence is increasingly viewed as better ensuring success for the organization.

Various methods of data collection are currently being used in different sectors by public, private and non-governmental organizations. Data collection using remotely controlled vehicles is common around the globe. Remotely controlled ground rovers, fitted with various types of sensors, are successfully patrolling a variety of locations, and are routinely used in response to hazardous incidents by public safety agencies. Remotely controlled watercraft have been deployed in harbors and ports, collecting information and monitoring the status of ships and materials containment. When employed by appropriately trained and practiced operators, remotely controlled vehicles could be invaluable during emergency response.

The integration of non-traditional vehicles into public safety applications is not without precedent. Bicycles and ATVs were initially considered to be only for recreation before being adapted into emergency management and public safety. Training programs and emergency scenarios were developed and implemented, turning these vehicles into widely accepted tools for public safety and other emergency management applications. The same process should be used for remotely controlled vehicles. The relatively low-cost for remote controlled vehicle acquisition and operation permits the simultaneous deployment of multiple types of vehicles that will provide expanded information gathering capability.

Aerial survey with conventional aircraft has long been used by public service agencies for both routine and emergency missions. The experimental use of remote controlled aircraft for mapping and search and rescue operations began around 2006. At the time, the Federal Aviation Administration (FAA) prohibited the use of remote control aircraft by public agencies unless a complex application process was successfully navigated. In September 2014, the FAA established the Section 333 exemption process, thereby permitting the commercial operation of small unmanned aircraft systems (UAS) in the private sector. The 333 exemption was the first step toward safely integrating unmanned aircraft into the nation's airspace. In response to this action, drone manufacturers accelerated the development of these small aircraft to make them a versatile tool. During their initial introduction into civilian public airspace, some reckless drone operations interfered with conventional aircraft operations, such as aerial firefighting. In order to better monitor small UAS activities, the FAA mandated that all small UAS had to be registered, and established civil and criminal penalties for non-compliance.

In August 2016, new FAA regulations were enacted for both public and private sectors of small UAS operations. Known as Part 107, these new regulations simplified the requirements for small UAS operations by public agencies, establishing flight rules and criteria for operator certification.

By employing small UAS, aerial data collection and its benefits are now accessible to organizations that cannot afford the costs of conventional aircraft operations. Small UAS can offer high-quality information collection at a fraction of the cost. They are not intended to replace conventional aircraft, only to provide supplemental coverage.

Time and effort must be invested by an organization in order to create a small UAS program that covers every step from equipment acquisition to data collection. Implementing a program may involve considerably long-range strategic planning and commitment of resources, both directly and indirectly, before a return can be realized. As such, the organization should consider the benefits of contracting with qualified and reputable providers to enhance their operations and produce a quicker return on investments.

SECTION 1

Introduction to Small UAS

A small unmanned aircraft system, also known as a small UAS, includes an unmanned aircraft, along with all of the supporting equipment needed for flight operations. The Federal Aviation Administration (FAA) states that a small unmanned aircraft must weigh 55 pounds or less.

There are two categories of small UAS: fixed-wing and rotary-wing. A new category, known as the transition, combines the features of both fixed and rotary-wing UAS.

Fixed-wing aircraft feature a rigid wing and a central fuselage. Lift is generated when the wing moves through the air. The propulsion system of a fixed-wing aircraft drives the propeller, creating forward speed. The increasing forward speed forces air over the wing and creates lift, enabling flight. A fixed-wing aircraft must remain in constant forward motion to maintain lift. Fixed-wing small UAS generally need a large area for takeoffs and landings.

Rotary-wing aircraft feature constantly rotating propellers that generate lift. Constant forward motion is not necessary for rotary-wing aircraft, hence their ability to hover. Rotary-wing aircraft can takeoff and land vertically, requiring a smaller area of operation for landing and takeoff. Rotary-wing aircraft can be configured as a helicopter (1 or 2 rotors), or as a multi-rotor (4 or more rotors).

Unmanned Aircraft Systems include the following components:

- Battery (Lithium-ion, Lithium Polymer)
- Electronic Speed Controller
- Motor
- Flight Controller & GPS
    - Gyroscopes
    - Accelerometers

- Magnetometer
- Barometric Pressure Sensor
- GPS Antenna and Receiver
- Servo or Actuator
- Ground Control Station
- Imaging/Data Sensor

Small UAS Definition

The Federal Aviation Administration (FAA) states that a small UAS includes an unmanned aircraft that weighs 55 pounds or less, and all of the support equipment (ground control, data monitors, etc.) necessary for its safe operation.

Today's small UAS platforms can offer situational awareness at a fraction of the operational cost of conventional helicopters and fixed-wing aircraft. While small UAS have significantly less payload capacity than conventional aircraft, they are able to carry modern small and lightweight cameras and sensors. Rapid deployment, modern sensors, longer flight times, and advanced data download capabilities make small UAS an invaluable resource in all situations.

The majority of small UAS fall into two general types: fixed-wing or rotary-wing. A new type, known as "transition" has been developed with the features of both fixed and rotary-wing UAS.

Fixed-Wing

Fixed-wing small UAS consist of a rigid wing with a preselected airfoil and a central fuselage. The propulsion system (electric, gas, or hybrid) drives the propeller of a fixed-wing UAS. As the motor turns the propeller, forward speed is generated by the thrust from the propeller. As the propeller increases the forward speed of the aircraft, air passes over the wing, generating lift and flight.

The lift generated by their wings gives fixed-wing small UAS the advantage of longer flight durations at higher speeds. This capability gives fixed-wing small UAS the ability to survey larger areas per given flight. Other advantages of fixed-wing small UAS include their stable flight characteristics and their ability to glide in the event of a power failure. Fixed-wing aircraft are able to carry greater payloads for longer flight times on less power, allowing the operator to carry larger sensors as well as multiple sensor configurations.

Fixed-wing small UAS tend to need a larger area for takeoffs and landings. While small fixed-wing small UAS may be hand-launched, larger ones need a catapult, bungee launcher, or even a runway for takeoff. Since fixed-wing aircraft require continuous airflow over their wings to generate lift, they must stay in constant forward motion and cannot remain stationary. Fixed-wing small UAS can be flown at low speeds over a specific area and remain aloft for extended periods of time, this is referred to as "loitering" over a point of interest.

Rotary-Wing

There are two primary types of rotary-wing aircraft-helicopters and multi-rotors. Helicopters have either one or two rotor blades. Most multi-rotors have four rotor blades but some may have six or eight.

The rotor blades on all rotary-wing aircraft are in constant movement. They produce the required airflow over their airfoil to generate lift. Since the rotor blades produce lift, constant forward movement is not required for flight.

The biggest advantage of rotary-wing small UAS is their ability to takeoff and land vertically, allowing for launch and recovery operations in more confined areas. The ability to hover and maneuver makes rotary-wing small UAS suitable for applications that require precision maneuvering and the ability to maintain a visual target for extended periods of time.

Rotary-wing small UAS generally have more complex mechanical and electrical systems than fixed-wing small UAS, meaning more involved maintenance and repairs. Rotary-wing aircraft generally have higher operational costs.

Due to the lower speeds and shorter endurance for rotary small UAS, the operator may require additional flights to survey significant areas, thereby facing higher operational costs.

Rotary-wing small UAS are popular because of their combined stability and vertical takeoff and landing (VTOL) capability. Rotary-wing small UAS generally have shorter endurance than fixed-wing small UAS, but their hovering capability makes them ideal for situational awareness. Often equipped with high-definition video cameras and video downlink, a rotary-wing small UAS can provide an instant eye in the sky.

Multi-Rotor

Multi-rotor small UAS generally use two or more pairs of identical fixed-pitched propellers; some turn clockwise (CW) and the others turn counter-clockwise (CCW). These propellers use independent speed variation on each rotor to achieve control. Changing the speed of each rotor allows the multi-rotor to:

- Specifically generate a desired level of lift.
- Locate the center of lift.
- Create a desired torque or turning force.

Multi-rotors offer solutions to some of the persistent problems in vertical flight. Torque-induced control issues can be eliminated by counter-rotation. The relatively small multi-rotor blades are easy to manufacture and maintain.

Multi-rotors require less maintenance than helicopters due to their mechanical simplicity. The smaller multi-rotor blades possess less kinetic energy, reducing their ability to cause damage and making the aircraft safer. Rotor blades on multi-rotors can be enclosed in shrouds, further reducing the potential for damage or personal injury.

The need for aircraft with greater maneuverability and improved hovering ability has led to a rise in multi-rotor research and development. Due to continued advancements in design, multi-rotors are becoming capable of complex autonomous missions that are currently impossible to conduct with other aircraft.

Characteristic	Fixed-Wing	Rotary-Wing	Multi-Rotor
Takeoff/Landing	Runway, hand launch or launcher	Vertical	Vertical
Maneuverability	Limited	Moderate	High
Maintenance	Lower	Higher	Lower

Transition

Some small UAS manufacturers are developing fixed wing aircraft with the ability to takeoff and land vertically. Typically, an electric VTOL system allows the transition from takeoff to forward flight. At the end of the flight, the aircraft slows and can land in the same confined space from which it launched. Vertical takeoff and landing is an essential part to operating small UAS in certain locations. The ability to launch a fixed wing aircraft from confined areas may provide some benefits. Allowing more flexibility and broader mission capability is desirable and those factors must be weighed against the additional complexity of a "transition" type of system.

Airframes & Components

Various materials are used for the fabrication of small UAS airframes. Fixed-wing small UAS are often manufactured from composite materials such as fiberglass, carbon fiber, or Kevlar®. Some fixed-wing airframes are made of molded foam reinforced with carbon fiber tubing.

The most common multi-rotor airframe is the "plate and boom" design. Center plates are fastened together and booms are attached symmetrically to support the motors and propellers. The airframes of some smaller multi-rotors consist of molded plastic shells.

All fixed-wing and multi-rotors small UAS share similar components. The primary components of small UAS are:

- Battery (Lithium-ion, Lithium Polymer)
- Electronic Speed Controller
- Motor
- Flight Controller & GPS
- Servo or Actuator
- Ground Control Station
- Imaging/Data Sensor

Lithium Battery

Most small UAS use a lithium battery to power the motor and other electrical components. Lithium batteries possess desirable qualities such as high energy density and the ability to be rapidly recharged. Lithium batteries do not require a periodic "deep discharge", and therefore are less likely to develop a charging "memory" common in other types of rechargeable batteries. A lithium battery pack powers an electrical bus that distributes the necessary power to each component. The electrical bus consists of a wiring harness that connects one or more voltage regulators that protect the smaller components from high voltage (power surges).

Lithium batteries are comprised of individual cells that are wired together to increase the voltage necessary for flight. Lithium batteries require proper handling and use, as mishandling them may cause a fire during charging or discharging.

Electronic Speed Controller

An electronic speed controller (ESC) is a circuit board that controls an electric motor's rotational speed. The ESC provides electronically generated three-phase electric power to the motor from the lithium battery. Most ESCs can be programmed to control low voltage cut-off limits, braking and direction of rotation. The ESC uses the proportional signal from the throttle to determine the amount of power required for the motor to maintain the desired speed.

ESCs are normally rated according to maximum current, for example 35 Amperes. ESCs with high ratings are larger and heavier than those with low ratings. The weight of an ESC may affect weight and balance calculations. Most modern ESCs contain a microcontroller (small computer) that interprets the input signal and appropriately controls the motor using built-in firmware.

Motor

While some small UAS may have gas or hybrid motors, almost all small UAS are powered by brushless electric motors. Brushless motors are popular because of their efficiency, power and longevity, compared to traditional brushed motors. Most small UAS use a type of brushless motor known as an outrunner. A brushless outrunner motor has an outer casing that rotates around its windings. Outrunners spin much slower than traditional electric motors but produce far more torque.

Flight Controller

Since no human is capable of controlling the rotational speeds of three or more motors simultaneously, a small computer is located on the multi-rotor to control it during flight. A flight controller (FC) is a small circuit board capable of being programmed to manage specific functions onboard the small UAS. Functions include managing the RPM of each motor or moving the flight control surfaces to direct the movement of the small UAS autonomously.

The term flight controller is often incorrectly used interchangeably with the word autopilot. An autopilot is a broader term and usually refers to the entire system used to control the trajectory of an aircraft without input by a human. Autopilots do not replace a human operator or pilot, but assist them in controlling the small UAS, allowing them to focus on other aspects of the flight operation.

All flight controllers have very small sensors that supplement their flight calculations. These tiny sensors are known as microelectromechanical systems, or MEMS. Often less than 1.0 mm in size, MEMS consist of a central unit that processes data (the microprocessor) and several components (microsensors) that interact with the surroundings. There are five basic types of microsensors used in flight controllers today:

- Gyroscopes
- Accelerometers
- Magnetometer
- Barometric Pressure Sensor
- GPS Antenna and Receiver

Gyroscopes sense the rotational movement about an axis. They are used for detecting changes in pitch, roll and yaw and provide stability during flight.

Accelerometers sense changes in acceleration forces in three axes. In addition to sensing changes in movement which is critical to flight stability, accelerometers can sense the force of gravity.

Magnetometers sense variations in the magnetic field of the earth. A magnetometer acts as an electronic compass for the flight controller which in turn, uses information from the magnetometer to determine its heading.

Barometric Pressure Sensors detect changes in altitude. An increase in altitude means a decrease in air pressure. Barometric pressure sensors can detect small increments in altitude changes. Information obtained from a pressure sensor can become inaccurate if there is a sudden change in weather in the immediate vicinity of the aircraft.

GPS Antenna and Receivers help determine the exact position (longitude and latitude) as well as altitude information. A GPS system allows small UAS to fly to exact points in space (waypoints). A multi-rotor with a GPS system can record its starting position and return to that location in case of an emergency.

Servos and Actuators

An actuator is a mechanism that converts energy into motion. An actuator can be hydraulic, pneumatic or electric. Its purpose is to move a flight control surface on a fixed-wing aircraft. Most small UAS use a type of actuator known as a servo-mechanism, or servo. A servo consists of a small electric motor that drives a train of reduction gears and a simple potentiometer position sensor that controls movement. The components are encased in a plastic or aluminum case and receive their input from the flight controller or the pilot. Electrical commands drive the motor to move the flight control surfaces such as the ailerons, rudder, and elevator.

Ground Control Station

The ground control station (GCS) provides operators command and control of the small UAS. Some GCS can display real-time video transmitted from the aircraft's cameras to personnel on the ground. The operator can capture screen images, store and play back data. A GSC facilitates real-time retransmission of video and metadata to an operations network.

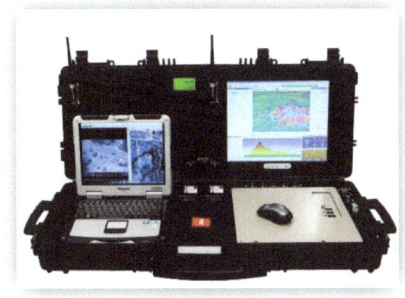

When embedded at remote locations, a GCS can be operated as a remote video terminal (RVT), enabling command centers or monitoring stations with the same viewing and analysis capability as the small UAS operator.

Some GCS are very compact, portable and can fit inside a small backpack. Advanced GCS give small UAS operators the option of manual or autonomous control to provide reconnaissance, surveillance and target acquisition.

Imaging Sensor

Modern small UAS are designed to carry a variety of cameras and sensors that can detect light, heat, chemicals, and even radiation. Some small UAS sensors are interchangeable for different applications while others are integrated components. Cameras and sensors are connected to the flight controller and other subsystems on the small UAS and on the ground.

SECTION 2

Federal Aviation Administration

The main responsibilities of Federal Aviation Administration are:

- Regulating air commerce to promote its development and safety
- Allocating the use of navigable airspace in the United States
- Regulating both civil and military operations in the National Airspace System (NAS)
- Developing and operating an Air Traffic Control (ATC) system
- Creating a system for registering civil aircraft
- Improving aircraft and flight safety through research and development

The Federal Aviation Regulations (FARs) govern aviation-related activities, and are promulgated in the Code of Federal Regulations — specifically Title 14. Activities regulated by Title 14 CFR include aircraft design and maintenance, pilot training activities, structure heights, obstruction lighting and marking, and operations of unmanned aircraft.

In 2012, the U.S. Congress authorized the FAA to safely integrate small UAS into the National Airspace System (NAS), thereby granting UAS operators increased and expedited access to the NAS. Sections 332, 333 and 334 were implemented specifically to address UAS operations while rules more specific to the regulation of small UAS were being developed.

In September 2016, the FAA enacted 14 CFR, Part 107 to regulate the operation of unmanned aircraft systems and the certification of remote pilots. Part 107 does not negate previous regulations, but rather offers alternative regulations under which a UAS may be operated.

The FAA's greatest interest is the safety of the general public and the promotion of voluntary compliance to the FARs through education. The FAA, however, retains the responsibility to enforce the regulations to better ensure the safety of the general public. State and local laws are being enacted to further regulate the safe operation of UAS. The FAA recognizes that state and local law enforcers may be in the best position to monitor and control unsafe UAS operations.

Background

On June 30, 1956, a Trans World Airlines (TWA) Super Constellation and a United Air Lines DC-7 collided over the Grand Canyon, Arizona, resulting in the loss of all 128 persons on board. The collision occurred while the aircraft were flying on a clear day in uncongested airspace. The accident dramatized the fact that, even though U.S. air traffic had more than doubled since the end of World War II, little had been done to mitigate the risk of midair collisions.

As a result of the 1956 accident, Congress established the Federal Aviation Administration (FAA), with the Federal Aviation Act of 1958 in order to increase aviation safety. The Federal Aviation Act transferred the Civil Aeronautics Authority's functions to a new and independent Federal Aviation Agency responsible for civil aviation safety.

In 1966, Congress authorized the creation of a cabinet department that would combine all major federal ground and air transportation responsibilities. This new Department of Transportation (DOT) began full operations on April 1, 1967. Accident investigation functions were then transferred to the National Transportation Safety Board (NTSB). The Federal Aviation Agency was renamed the Federal Aviation Administration and its main responsibilities now include:

- Regulating air commerce to promote its development and safety
- Allocating the use of navigable airspace in the United States
- Regulating both civil and military operations in the National Airspace System (NAS)
- Developing and operating an Air Traffic Control (ATC) system
- Creating a system for registering civil aircraft
- Improving aircraft and flight safety through research and development

The primary responsibility of the FAA is aviation safety regulation. The FAA promulgates rules, regulations, and minimum standards relating to the manufacture, operation and maintenance of aircraft. In the interest of safety, the FAA rates and certifies people working on aircraft, including medical personnel. The FAA also regulates airports based on the services provided. For example, airports that serve air carriers are required to maintain an Airport Operating Certificate. The FAA also provides standards for the design of heliports and airports.

The FAA has significant enforcement power, including the ability to delay flights or ground aircraft deemed non-airworthy and to suspend and/or revoke the license of pilots and other flight personnel who violate FAA regulations.

The FAA is authorized to test and evaluate aviation systems, materials and procedures at any phase in their development. The agency may assign independent testing at key decision points in the development cycle of these elements.

Federal Aviation Regulations

Federal Aviation Regulations (FARs) are regulations that govern aviation-related activities in the United States. The FARs are part of Title 14 of the Code of Federal Regulations (CFR). Activities regulated include aircraft design and maintenance, pilot training activities, structure heights, obstruction lighting and marking, and unmanned aircraft operations. The rules are designed to promote safe aviation, protect pilots, flight crew, passengers and the general public from unnecessary risk.

The FAA defines an unmanned aircraft as "aircraft operated without direct human intervention from within the aircraft." Because small UAS are aircraft, they are subject to all applicable FAA regulations, including the statutory requirements regarding registration set forth in Title 49 United States Code (U.S.C.) 44101(a), and further prescribed in 14 CFR part 47 and Part 48.

Federal law requires that a person may only operate an aircraft when it is registered with the FAA.

FAA Modernization and Reform Act of 2012

In 2012, Congress authorized the FAA to safely integrate small UAS into the NAS. There are several sections of this federal law that grant small UAS operators increased and expedited access to the NAS.

Section 332. Integration of civil UAS into the NAS.

The FAA must draft new regulations for small UAS that will allow for civil operation of such systems in the NAS. The FAA must develop a comprehensive plan to safely accelerate the integration of civil UAS into the NAS.

Section 333. Special rules for certain unmanned aircraft systems.

The FAA shall determine if certain UAS may operate safely in the NAS before completion of the plan and proposed rulemaking as required by section 332 of this Act.

If the FAA determines under this section that certain small UAS may operate safely in the NAS, requirements for the safe operation of such aircraft systems shall be established.

An assessment shall be made as to the types of small UAS that will not create a hazard to users of the NAS. Those small UAS that meet the safety criteria shall be granted an "exemption" from the current regulations.

Section 334. Public unmanned aircraft systems.

The FAA shall issue guidance regarding the operation of public small UAS to:

- Expedite the certificate of authorization (COA) process
- Provide guidance on a public agency's responsibility when operating a small UAS without a civil airworthiness certificate issued by the Administration
- Develop and implement operational and certification requirements for the operation of public small UAS
- Simplify the process for issuing certificates of waiver or authorization with respect to applications seeking authorization to operate public small UAS.

Part 107 Regulations

As mandated by Congress, the FAA has amended its regulations to allow the civil operation of small unmanned aircraft systems in the NAS under Title 14 Code of Federal Regulations (14 CFR) Part 107. Effective on August 28th, 2016, these regulations address the operation of unmanned aircraft systems and certification of remote pilots. Part 107 allows small UAS operations for many different non-recreational purposes without requiring airworthiness certification.

The FAA has previously accommodated non-recreational small UAS use through various mechanisms, such as special airworthiness certificates, Section 333 exemptions, and certificates of waiver or authorization (COAs). Part 107 is the next phase of integrating small UAS into the NAS. Part 107 allows for routine civil operation of small UAS in the NAS and provides safety rules for those operations. The rulemaking also establishes a new class of airman certificate tailored to remote pilots of small UAS.

Under part 107, a public entity can continue to operate small UAS under a COA or operate in compliance with the new rule. Part 107 will provide greater flexibility to public aircraft operations because it allows small UAS public aircraft operations to voluntarily opt into the part 107 framework. In other words, a government entity may now operate a small UAS as a civil rather than a public aircraft and comply with part 107 requirements instead of obtaining a COA from the FAA.

FAA Enforcement

The FAA promotes voluntary compliance by educating individual small UAS operators about safe operations under current regulations and laws. The FAA has a number of enforcement tools available including warning notices, letters of correction and civil penalties. The FAA may take enforcement action against anyone who conducts unauthorized small UAS operations or operates in a way that endangers the safety of the NAS.

Regardless of the type of small UAS operation, the FAA's statutes and the Federal Aviation Regulations (FARs) prohibit any conduct that endangers persons and property on the ground, or endangers the safe operation of other aircraft in the NAS. Many state and local governments are enacting their own laws regarding the operation of small UAS. All small UAS operators must also be aware of state and local laws, as well as broadly applicable laws such as assault, criminal trespass, or injury to persons or property.

The FAA retains the responsibility for enforcing FARs including those applicable to the use of small UAS. The FAA does recognize, however, that a state and local law enforcement agencies (LEA) is often in the best position to deter, detect, immediately investigate and stop unauthorized or unsafe small UAS operations.

SECTION 3

National Airspace System

Understanding the classification of airspace prior to planning small UAS operations is critical to safe and legal operations. The FAA may designate or change the airspace and requirements for small UAS operations at any time due to an emergency or security situation.

There are two types of aeronautical charts that assist in flight planning; Sectional Aeronautical Charts and Terminal Area Charts. Sectional charts are the primary navigational charts used for visual flight. They include topographic information and visual references such as highways, populated areas and other landmarks.

Terminal Area Charts are intended for aircraft operations near the nation's busiest airports and feature more detailed information than sectional charts.

The National Airspace System (NAS) includes the airspace, navigation facilities and airports of the United States along with their associated information, services, rules, regulations, policies, procedures, personnel and equipment. There are four types of airspace within the NAS:

- Controlled
- Uncontrolled
- Special Use
- Other Airspace Areas

Controlled airspace is a generic term that covers the different classifications of airspace and defined dimensions (vertical or lateral boundaries) within which Air Traffic Control (ATC) service is provided. Controlled airspace consists of Class A, B, C, D and E. Uncontrolled airspace - Class G airspace is the portion of the airspace that has not been designated as Class A, B, C, D, or E.

Special use airspace is designated airspace in which certain activities must be confined, or where limitations are imposed on all aircraft operations that are not part of those activities.

- Prohibited areas
- Restricted areas
- Warning areas
- Military Operation Areas (MOAs)
- Alert areas
- Controlled Firing Areas (CFAs)

"Other airspace areas" is a general term referring to other airspace areas such as a Temporary Flight Restriction (TFR) and National Security Area (NSA).

Aeronautical Charts

Pilots and operators use aeronautical charts, published by the FAA, for flight planning and situational awareness. Aeronautical charts contain useful information such as air traffic frequencies, obstacle clearances and airspace boundaries. Aeronautical information changes rapidly, so it is essential that pilots use current publications. Numerous websites and apps now provide access to this information but it is important to understand the origin of this information. The two main types of aeronautical charts are the Sectional Aeronautical Chart and the Terminal Area Chart (TAC).

Sectional Aeronautical Charts are the primary navigational charts used for visual flights and are scaled to 1:500,000. Sectional charts include topographic information and visual references such as highways, populated areas, and other landmarks. The aeronautical information on sectional charts includes visual and radio aids to navigation, airports, controlled airspace, restricted areas, obstructions and related data. These charts are updated every six months.

Terminal Area Charts (TACs) are intended for aircraft operations near the nation's busiest airports. They provide a large-scale, detailed portrayal of selected metropolitan areas such as Los Angeles, Chicago and New York. The scale on a TAC is 1:250,000. Because of the larger scale, Terminal Area Charts feature much more detailed information than sectional charts. These charts are updated every six months.

National Airspace System (NAS)

The National Airspace System (NAS) was created for civil aviation in order to get aircraft from one airport to another in a safe and efficient manner. The NAS includes the airspace, navigation facilities and airports of the United States along with their associated information, services, rules, regulations, policies, procedures, personnel and equipment. There are four types of airspace within the NAS:

- Controlled
- Uncontrolled
- Special Use
- Other Airspace Areas

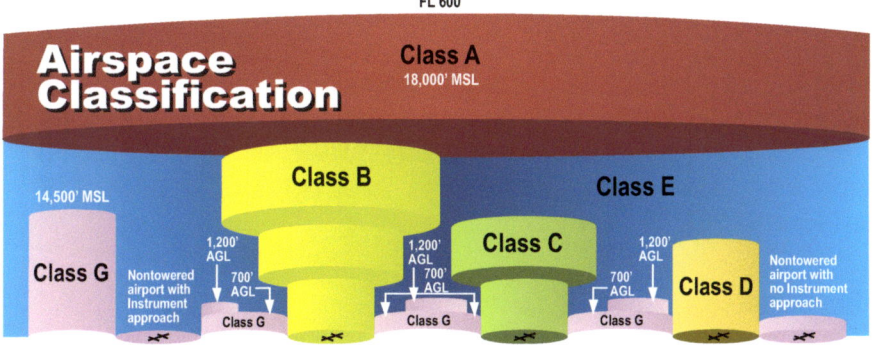

Controlled Airspace

Controlled airspace is a generic term that covers the different classifications of airspace and defined dimensions (vertical and lateral boundaries) within which Air Traffic Control (ATC) service is provided. Each type of airspace has its own required level of air traffic control services and its own minimum operational requirements for all aircraft. Controlled airspace consists of Class A, B, C, D and E.

Class A airspace covers the entire United States and extends between 18,000 feet and 60,000 feet above mean sea level (MSL). Class A airspace is primarily used by jets and airliners traveling long distances between major cities.

Class B airspace extends from the surface to 10,000 feet above MSL feet and most often has a circular diameter of 40 nautical miles. Class B airspace surrounds the nation's busiest airports and airport hubs in cities like Boston, Chicago and Los Angeles. Class B airspace is designed to help manage the flow of high volumes of airline traffic as these aircraft descend from high-altitude flight levels into the lower altitudes and eventually into the airport itself. The configuration of each Class B airspace area is individually tailored to the airport in a tiered fashion and resembles an upside-down wedding cake to help safely direct aircraft in and out of the main airport.

Class C airspace extends from the surface to 4,000 feet above ground level (AGL), with a circular diameter of 20 nautical miles. Class C airspace surrounds airports that have an operational control tower, are serviced by a radar approach control and have a certain number of commercial passenger aircraft operations.

Class D airspace is generally airspace from the surface to 2,500 feet AGL surrounding those airports that have an operational control tower. Class D airspace has a circular diameter of five nautical miles.

Class E airspace is any controlled airspace other than Class A, B, C or D. Class E airspace extends upward from either the surface or a designated altitude to the overlying or adjacent controlled airspace. Unless designated at a lower altitude, Class E airspace begins at 14,500 feet above MSL over the United States, up to but not including 18,000 feet above MSL.

Uncontrolled Airspace

Class G airspace is the portion of the airspace that has not been designated as Class A, B, C, D, or E. It is therefore designated as uncontrolled airspace. Class G airspace extends from the surface to the base of the overlying Class E airspace. Although ATC has no authority or responsibility to control air traffic, small UAS operators should remember there are FAA rules and regulations that apply to Class G airspace. Even though Class G airspace is uncontrolled, that does not mean it is unregulated.

This chart identifies the altitude for each class of airspace. Among these, Class G is the only non-controlled airspace. Prior permission from Air Traffic Control is required to operate in Class B, C, or D airspace, or within the lateral boundaries of the surface area of Class E airspace designated for an airport.

Special Use Airspace

Special use airspace is designated airspace in which certain activities must be confined, or where limitations are imposed on all aircraft operations that are not part of those activities.

These limitations are often due to military use of the area or national security concerns.

The special use airspace depicted on aeronautical charts includes the area name or number, effective altitude, time and weather conditions of operation. Special use airspace usually consists of:

- Prohibited areas
- Restricted areas
- Warning areas
- Military Operation Areas (MOAs)
- Alert areas
- Controlled Firing Areas (CFAs)

A **Prohibited Area** contains airspace of defined dimensions within which aircraft flight is prohibited. Such an area is established for national security or other reasons associated with national welfare. They are published in the Federal Register and are depicted on aeronautical charts. Prohibited areas are indicated on aeronautical charts with a "P" followed by a number (e.g., P-49). Examples of prohibited areas include Camp David and the Washington National Mall in D.C. where the White House and Congressional buildings are located.

A **Restricted Area** is airspace designated for use by the military in which the local controlling authorities have determined that air traffic must be restricted for safety or security concerns. It is one of many types of special use airspace designations and is depicted on aeronautical charts with the letter "R" followed by a serial number. Restricted areas denote the existence of unusual, often invisible, hazards to aircraft (e.g., artillery firing, aerial gunnery, or guided missiles). Penetration of restricted areas without authorization from the using or controlling agency may be extremely hazardous.

Restricted areas are charted with an "R" followed by a number (e.g., R-4401) and are depicted on the aeronautical chart appropriate for use at the altitude being flown. Restricted area information can be obtained on the back of the chart.

Warning Areas are similar in nature to restricted areas; however, the United States government does not have sole jurisdiction over the airspace. A warning area is airspace of defined dimensions, extending from 12 miles outward from the coast of the United States and containing activity that may be hazardous to nonparticipating aircraft. The purpose of such areas is to warn nonparticipating pilots of the potential danger. A warning area may be located over domestic or international waters or both. The airspace is designated with a "W" followed by a number (e.g., W-237).

Military Operation Areas consist of airspace with defined vertical and lateral limits established for the purpose of separating certain military training activities from civilian air traffic.

MOAs are depicted on a sectional flight chart, and are named rather than numbered (e.g., "Camden Ridge MOA"). MOAs are further defined on the back of the sectional charts with times of operation, altitudes affected, and the controlling agency.

Alert Areas consist of airspace with defined vertical and lateral limits established for the purpose of separating certain military training activities from civilian air traffic.

MOAs are depicted on a sectional flight chart, and are named rather than numbered (e.g., "Camden Ridge MOA"). MOAs are further defined on the back of the sectional charts with times of operation, altitudes affected, and the controlling agency.

Controlled Firing Areas contain activities, which, if not conducted in a controlled environment, can be hazardous to nonparticipating aircraft. The difference between CFAs and other special use airspace is that activities must be suspended when a spotter aircraft, radar, or ground lookout position indicates an aircraft might be approaching the area. There is no need to chart CFAs since they do not cause a nonparticipating aircraft to change its flight path.

Other Airspace Areas

"Other airspace areas" is a general term referring to the majority of the remaining airspace. Examples of other airspace areas include Temporary Flight Restriction (TFR) and National Security Area (NSA).

A **Temporary Flight Restriction (TFR)** is a regulatory action issued by the FAA to restrict certain aircraft from operating within a defined area, on a temporary basis, to protect persons or property in the air or on the ground. A TFR may be requested by various entities including federal agencies; regional directors of the Office of Emergency Planning, Civil Defense State Directors and civil authorities directing organized relief air operations such as the Office of Emergency Planning.

A TFR is disseminated by the Notice to Airmen (NOTAM) system. The NOTAM begins with the phrase "FLIGHT RESTRICTIONS" followed by the location of the temporary restriction, effective time period, area defined in statute miles, and altitudes affected. The NOTAM also contains the FAA coordination facility and telephone number, the reason for the restriction, and any other information deemed appropriate.

The small UAS operator must check the NOTAMs as part of flight planning. Some of the purposes for establishing a TFR are:

- Protect persons and property in the air or on the ground during incidents involving hazardous materials, fire or smoke.
- Provide a safe environment for the operation of disaster relief aircraft.
- Prevent an unsafe congestion of sightseeing aircraft above an incident or event, which may generate a high degree of public interest.
- Protect declared national disasters for humanitarian reasons.
- Protect the President, Vice President, or other public figures.
- Provide a safe environment for space agency operations.

A **National Security Areas (NSA)** consists of airspace of defined vertical and lateral dimensions established at locations with a requirement for increased security and safety of ground facilities. Examples include sporting events (Super Bowl), political conventions, parades, etc.

Flights in NSAs may be temporarily prohibited by regulation under the provisions of Title 14 of the Code of Federal Regulations (14 CFR) part 99. Flight restrictions are disseminated via NOTAM.

SECTION 4

Types of Small UAS Operations

The FAA has developed rules and regulations for small UAS operations depending on the classification of the entity and the type of operation being conducted. The following types of small UAS operations are identified:

- Public Operations (Government)
- Civil Operations (Non-Government)
  - Commercial (Non-Recreational)
  - Model Aircraft (Recreational)

Public Operations (Government)

A Public Aircraft Operation (PAO) is limited by federal statute to certain government operations within U.S. airspace. An aircraft must meet at least one of the following criteria to be considered a "public aircraft."

1. An aircraft used only for the United States Government;
2. An aircraft owned by the Government and operated by any person for purposes related to crew training, equipment development, or demonstration;
3. An aircraft owned and operated by the government of a State, the District of Columbia, or a territory or possession of the United States or a political subdivision of one of these governments; or
4. An aircraft exclusively leased for at least 90 continuous days by the government of a State, the District of Columbia, or a territory or possession of the United States or a political subdivision of one of these governments. There is an exception to the 90-day rule but only for search and rescue operations fulfilling certain requirements.

Once the definition of public aircraft is established, the operation must be analyzed to ensure it fulfills a government function such as national defense, intelligence missions, firefighting, search and rescue, law enforcement, etc. In addition, the government agency may not receive any type of reimbursement for the operation. If in doubt, the FAA should be consulted for an interpretation.

Government agencies and public entities intending to conduct public small UAS operations must comply with the following requirements:

- **Declaration Letter** – A letter from the city, county, or state attorney's office assuring the FAA that the proponent is recognized as a political subdivision of the government of the state under Title 49 of the United States Code (USC) section (§) 40102(a)(41)(c) or (d) and that the proponent will operate its unmanned aircraft in accordance with 49 USC. § 40125(b) (not for commercial purposes).
- **FAA issued Certificate of Waiver or Authorization (COA)** – A COA is a waiver issued by the Air Traffic Organization (ATO) to a public operator for a specific small UAS activity. After a complete application is submitted, the FAA conducts a comprehensive operational and technical review. If necessary, provisions or limitations may be imposed as part of the approval to ensure the small UAS can operate safely with other airspace users. In most cases, the FAA will provide a formal response within 60 days after a completed application is submitted.

Civil Aircraft Operations

Any operation that does not meet the statutory criteria for a public aircraft operation is considered a Civil Aircraft Operation (CAO) and must be conducted in accordance with all FAA regulations applicable to the operation. Individuals who fly model aircraft for hobby or recreational purposes do not need authorization from the FAA. All other civil small UAS operations in the national airspace require FAA authorization. There are presently three (3) methods for receiving FAA authorization to fly civil (non-governmental) small UAS:

- Title 14 Code of Federal Regulations (14 CFR), Part 107 – The Federal Aviation Administration (FAA) has amended its regulations to allow the operation of small unmanned aircraft systems (UAS) in the National Airspace System. Part 107 allows for routine civil operations of small UAS and provides safety rules for those operations. Small UAS civil operations may only be conducted by eligible persons with a remote pilot certificate.
- Section 333 Exemption – This authority is being leveraged to grant case-

by-case authorization for certain unmanned aircraft to perform commercial operations. The Section 333 exemption process authorizes operators who can demonstrate they can provide a level of safety equivalent to flying conventional aircraft. The operators are exempt from rigorous requirements for certifying their small UAS.

- Special Airworthiness Certificate (SAC) – In certain circumstances, a manufacturer may apply for a special airworthiness certificate for a small UAS. The SAC would be granted in the experimental category and only be applicable for the purposes of research and development, crew training, and market survey. Commercial operations are not permitted with experimental aircraft. Aircraft with a SAC have strict limitations for their operation as described below.

Experimental aircraft test new design concepts, aircraft equipment, installations, or new uses for aircraft. Crew training is limited to the applicant's flight crews, which would be the manufacturer's employees trained for that experimental aircraft. Market survey operations must be specified in advance with the estimated number of flights required as well as the flight area requested.

Model Aircraft (Hobby or Recreational)

Model Aircraft Operations are for hobby or recreational purposes only. The owner must register the aircraft if it weighs between 0.55 and 55 lbs. and follow the safety guidelines established by any nationwide community-based organization.

Model aircraft must be operated within all of the following parameters:

- Strictly for hobby or recreational use
- Must give way to all other aircraft
- Flown within visual line of sight of the operator
- Must weigh less than 55 lbs. unless certified through design, construction and inspection by community based organization
- Must be operated in accordance with community based set of safety guidelines
- If within five statute miles of airport, the operator must notify the airport operator and control tower (if towered)
- Registration and markings requirement must be met
- Must be flown below 400 feet above ground level at all times

A Public Aircraft Operation (PAO) is limited by federal statute to certain government operations within U.S. airspace. Once the definition of public aircraft is established, the operation must be analyzed to ensure it fulfills a government function. A government agency may not receive any type of reimbursement for the operation and must comply with the following requirements:

- Declaration Letter
- FAA issued Certificate of Waiver

Any non-recreational or hobby operation that does not meet the statutory criteria for a public aircraft operation is considered a Civil Aircraft Operation (CAO). Civil Aircraft Operations must receive FAA authorization to operate in the National Airspace System. The three methods for receiving FAA authorization are:

- Title 14 Code of Federal Regulations (14 CFR), Part 107
- Section 333 Exemption
- Special Airworthiness Certificate (SAC)

Model Aircraft Operations are for hobby or recreational purposes only. The owner must register the aircraft and follow safety guidelines established by the FAA.

SECTION 5

Small UAS for Emergency Management

Small unmanned aircraft systems (UAS) can be an effective alternative to conventional aircraft for emergency missions that are managed under the National Incident Management System and Incident Command System (NIMS-ICS). Small UAS can be deployed almost immediately and can collect real time information in emergency situations.

The integration of small UAS can be achieved seamlessly into the Common Operational Picture (COP), which is the core Department of Homeland Security (DHS) situational awareness capability for effective decision-making, rapid staff actions, and appropriate mission execution. Small UAS can rapidly provide first responders with the information that gives them the ability to quickly assess and properly respond to emergencies.

 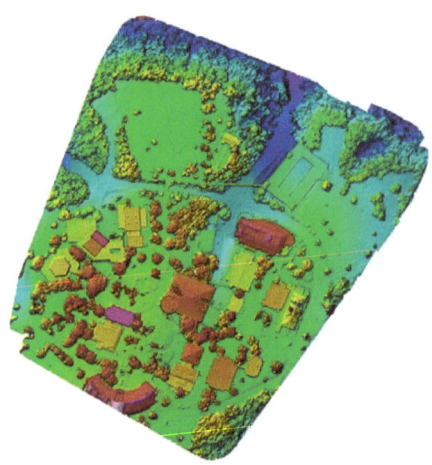

New data-processing software can now produce 2D and 3D maps and models from lightweight digital cameras. Smaller, lighter and more capable sensors can now be attached to small UAS to capture high-quality imaging in a real-time environment.

Currently available sensor technology includes:

- Thermal Imaging
- Hyperspectral Imaging
- Multispectral Imaging
- Light Imaging, Detecting and Radar (LIDAR)
- Radar Imaging
- Interferometric Synthetic Aperture Radar (InSAR)

Collected data can be used to guide emergency planners and incident commanders response teams in the preparation, mitigation, response and recovery efforts. Small UAS can be beneficial in the following areas:

- Damage Assessment
- HazMat Response
- Wildfire Operations
- Structure Fire Operations
- Search and Rescue
- Law Enforcement Support
- Transportation Infrastructure
- Disaster Planning and Response
- Environmental Analysis
- Agricultural Analysis
- Humanitarian Assistance

Small UAS Integration

The main goal of integrating small unmanned aircraft systems into missions that are managed under the National Incident Management System and Incident Command System (NIMS-ICS) is to collect data remotely in a safe and efficient manner. Today's small UAS can simultaneously provide real time information to Incident Command Posts (ICPs), Emergency Operations Centers (EOCs), and other stakeholder organizations. Small UAS can be deployed immediately, staged near the affected area, and operated at various altitudes to monitor and document multiple incident-related actions. The variety of commercially available small UAS platforms and sensor technologies offers a wide range of applications in both emergency and non- emergency operations in a variety of applications and disciplines.

The integration of small UAS resources into the emergency management mission that are managed under NIMS/ICS can be done seamlessly. Small UAS offer an additional method to collect, analyze and communicate data using the existing NIMS-ICS characteristics of effective communication systems:

- Interoperable – able to communicate within and across agencies and jurisdictions
- Reliable – able to function in the context of most any type of emergency
- Portable – built on standardized radio technologies, protocols and frequencies
- Scalable – suitable for use on a small or large scale incident
- Resilient – perform despite damaged or lost infrastructure
- Redundant – use alternate methods of communication when primary systems are unavailable

Common Operational Picture

Common Operational Picture (COP) is the core Department of Homeland Security (DHS) situational awareness capability for effective decision-making, rapid staff actions, and appropriate mission execution. Information collected with small UAS can provide first responders with the ability to quickly assess the complexity of an emergency and the knowledge required to respond accordingly. Immediate responsiveness relies on the quality, accuracy, timeliness and usability of the collected information. Small UAS are capable of adapting to the dynamic changes that occur in emergency situations. Rapid information exchange and efficient collaboration between emergency services agencies are essential to the safety of first responders as well as the public they serve.

Small UAS Sensor Technology

As technology becomes smaller, simpler, lighter and more capable, it offers users an ever-increasing degree of mission flexibility.

Compact imaging sensors enhance the mission capabilities of small UAS because of their relatively low weight and small size. Reducing the weight of one sensor makes it possible to install additional sensors and potentially increase flight times. The adaptability of various sensor types allows small UAS to meet numerous mission profiles.

High-definition imagery translates into greater mission capability. An image with higher resolution provides more information and better picture quality. More image information and better quality may result in more ground being covered at a safer distance and with less risk to the platform and personnel. New data-processing software can now produce 2D and 3D maps and models from lightweight digital cameras.

Small UAS can provide live video streaming to multiple users for fast dissemination. Data for numerous applications can be gathered simultaneously by a variety of small UAS sensor technologies. The following is a list of sensor technologies currently available for small UAS.

Thermal Imaging is a technique that improves the visibility of an object in a dark environment by detecting that object's infrared radiation (heat) and creating an image based on that information. Forward looking infrared (FLIR) cameras are a type of thermal sensor that can be used for real-time imaging.

Hyperspectral Imaging collects and processes information from across the electromagnetic spectrum. The goal of hyperspectral imaging is to obtain the spectrum for each pixel in the image of a scene, with the purpose of finding objects and identifying materials.

Multispectral Imaging captures image data at specific frequencies across the electromagnetic spectrum. Multispectral imaging is used to monitor the progress of wildfires, detect chemical irregularities such as asbestos and detect thermal anomalies in water in volcanic regions.

Light Imaging, Detecting and Ranging (LIDAR) is a remote sensing method which uses pulsed laser light to measure distances to the Earth, creating 3D images that can be used for decision making processes across many sectors. LIDAR sensors can penetrate dense vegetation and capture structures that satellites cannot see.

Radar Imaging uses an echo measuring system to create two-dimensional images. It can be operated despite the presence of obstacles that obscure the target such as sand, water, or walls. Radar imaging helps predict, track and respond to natural hazards by monitoring surface deformation.

Interferometric Synthetic Aperture Radar (InSAR) uses two or more synthetic aperture radar images to generate maps of surface deformation. The technique can measure structural stability after disasters.

Small UAS Emergency Management Applications

The use of small UAS can be beneficial to the following areas:

- **Damage Assessment:** Assess and document the level of structural damage across all types of critical infrastructure (e.g. power lines, bridges, residential, and commercial infrastructure). Initially, Emergency Management and Public Safety agencies rely on the publics reports of damages to plan for and provide the appropriate post-disaster response. Subsequently, resources such as volunteers may be trained to perform Preliminary Damage Assessments (PDAs) and locate areas in need. For example, the state of Kentucky has modeled a PDA of critical infrastructure nodes referred to as a SWEATT-MS report which is a rapid analysis of:
    - (S) Sewer – waste
    - (W) Water – potable
    - (E) Electric
    - (A) Academics – all educational, both public and private facilities
    - (T) Transportation – roads, bridges, rail
    - (T) Telecommunications Technology – wireless, satellite
    - (M) Medical – all facilities clinics and hospitals
    - (S) Security

    The SWEATT-MS report is then coded Red, Yellow, or Green. Red indicates that the element is damaged and unusable. Yellow indicates damage, but still may be of use. Green means little to no damage or impact. Conducting this assessment and analysis may be challenging due to the damage caused by the event, and the limited access to the areas of impact. Aerial surveys by piloted aircraft can be expensive and time consuming to arrange, as they require prior approval for reimbursement. A small UAS can be rapidly deployed to assess the areas without the need for prior approval and would be more cost effective.

- **HazMat Response:** Assist in detecting and establishing the type, plume, concentration and direction of hazardous chemical releases. They can also help determine Exclusion (Hot) and Support (Cold) Zones, guide in planning for Shelter in Place vs. Evacuation, and related response planning. Aerial surveys by piloted aircraft may not be practical in assessing a Hazardous Materials incident as the pilot and observer may be exposed to the material being surveyed. Title 29 CFR 1910 requires the use of Level A (fully encapsulated suit) protection when personnel may be exposed to undetermined material concentrations until safe exposure levels are established. Small UAS can be used to perform safer and more practical aerial assessments.

- **Wildfire Operations:** Help first responders assess the origin, progression and extent of a wildfire. Locating the fire and mapping its margins can be a challenge for firefighters due to smoke and rapidly changing wind and weather conditions. Small UAS can help detect hot spots when used to map fire zones. When deployed in coordination with an Air Attack, small UAS can guide firefighting aircraft in dispensing fire retardant, thereby reducing the operating costs. Depending on the retardant being dropped – water vs. chemical – the cost to operate full-size firefighting aircraft can be as much as $5,000.00 per drop. U.S. Department of the Interior policy states that wildfires on Federally-controlled lands and National Parks are used to control and eliminate the accumulation of fuel buildup and decrease the likelihood of serious hotter fires. As opposed to a fire suppression management operations, the mission transitions to a fire use management operation, and the protection of structures while "guiding" the fire away from development to less vulnerable areas. Small UAS can assist in reviewing and updating incident maps in the Wildland Urban Interface, aid in identifying escape routes and safety zones, and act as an unmanned, auxiliary "Look Out."

- **Structure Fire Operations:** Inspect buildings that have been exposed to fire. Determining the margins and progression of a structure fire may prove challenging and dangerous for firefighters on the ground. Small UAS can move closer and assist with identifying and locating occupants inside the building. Small UAS can help identify and map escape and rescue routes.

- **Search and Rescue:** Assist in search efforts following earthquakes, floods and tornadoes, as well as search for missing individuals in remote areas, and in bodies of water. Small UAS can be equipped with Forward Looking Infrared (FLIR) sensors to locate injured or missing persons during periods of darkness by measuring temperature differences.

- **Law Enforcement Support, Security, and Surveillance:** Provide support to emergency responders for various incidents such as barricaded persons, hostage situations, crowd control, civil unrest, active assailant and surveillance of critical locations during and post incident. Small UAS can perform surveillance operations of remote property boundaries, physical access points and critical infrastructure nodes such as electrical substations, pipelines and pump stations.

- **Transportation Infrastructure:** Inspect earthquake damage, traffic accidents, derailments, mass evacuations, bridge collapses, dams and reservoirs, route analysis for special events and VIP emergency ingress and egress.

- **Disaster Planning and Response:** Provide geophysical analysis for measuring the impact of hurricanes, earthquakes, floods, volcanoes, avalanches and mudslides, coastal flooding, storm surge, and tsunamis.
- **Environmental Analysis:** Assist in retrieving air and water samples for quality analysis to determine possible threats to public safety.
- **Agricultural Analysis:** Assist in analyzing the condition of crops and other critical food sources in the event of a human or natural disaster. They can also be used to monitor the impact of weather on animal production operations over a large area despite inclement weather conditions (snow or water accumulation). In areas with limited government services, private land owners rely on hiring and deploying resources to patrol large tracts of land and monitor for fires. Small UAS can greatly reduce the number of resources required for monitoring large areas, thus saving money for large timber operations.
- **Humanitarian Assistance and Disaster Relief Operations:** Provide technical support to federal, military, state, non-governmental organizations, and faith-based organizations during humanitarian assistance and worldwide disaster relief operations. Seasonal workers and immigrant populations can be among the most vulnerable and least informed during a disaster. Because they may already reside in substandard conditions, they may also have the greatest needs when recovering from an incident. Due to mistrust of government, language and cultural barriers, these populations can represent a greater challenge when it comes to assistance during disaster recovery. Identifying gathering locations with a small UAS is essential to providing assistance to a potentially underserved population.
- **Public Health Crisis:** Securely transport critical medications when hospitals and healthcare facilities are unable to use normal modes of transportation due to a disaster.
- **Special Event Planning and Operations:** Assist in collecting route analysis information and developing ingress and egress routes for presidential visits, sporting events and major parades.
- **Emergency Communications:** Provide Wi-Fi and cellular phone relay services in remote wilderness areas during emergency operations.
- **Data for Post-Mission Analysis Activities:** Collect data for "forensic" support through case studies and incident simulation for various types of disaster incidents. Emergency Management planning, training, exercising and mitigation, are all based on lessons learned from previous, similar experiences. Post incident analysis and "After Action Reports" are routinely utilized by many jurisdictions for planning, training and exercising.

- **Preparedness and Response Exercise Support:** Add to the realism of training by providing imagery without the time and cost burden of actual deployment of personnel and equipment during a full-scale exercise.

- **Critical Infrastructure Key Resource Analysis Support:** Assist onsite personnel in conducting threat and risk assessments. Small UAS can collect still and video imagery of remote areas such as outer perimeters, inhospitable environments and areas with hazardous material storage. Additionally, small UAS can be used to survey roof tops, power plants, chemical plants, storage areas, fuels depots, dams and reservoirs.

- **Geographic Information Systems (GIS) Support:** Collect and displays GIS data in "layers" to provide a visual representation of surface information in the format and accuracy requested by the user. Recently collected data can be compared with archived data to provide real time information (e.g., snow accumulation or water accumulation). In certain instances, data obtained from archived resources and established plat maps may not provide the detail to be considered useful in the field. For instance, a large acreage may contain one or more dwellings. The plat map may only have a single symbol within the margin of the property lines indicated on the map, indicating the presence of occupancy, but not indicating the actual location, number, or type of occupancy. Before the map can be updated with any accuracy, many hours are spent in "ground truthing" the plat map, and updating the database. Small UAS can save a considerable amount of time by identifying points, relaying images and using geolocation technology to determine and transmit the exact coordinates for those points. This information would be invaluable in helping identify areas vulnerable to a variety of threats. In the case of wildfire, chemical fire retardant is routinely dispersed from aircraft on buildings in remote areas that are deemed to be in harm's way. The buildings must be located, identified and their coordinates plotted.

Cost Benefits

According to the FAA, small UAS can create significant cost savings for public entities and government agencies that may be currently using conventional aircraft for these operations. The FAA expects that some of the larger public agencies would train their own operators and purchase and operate their own small UAS.

The majority of the smaller public sector entities that cannot afford to train their personnel to fly a small UAS would most likely rely on Mutual Aid Agreements and/or contract these services out to commercial small UAS private sector enterprises as needed.

Community Engagement and Safety of the Public

Public sector entities interested in integrating small UAS technology into emergency response operations are encouraged to educate the local community. Transparency is essential when it comes to reassuring the community that the operator is complying with local, state, and federal laws, rules, statutes and regulations.

The safety of the public must be a top priority for all entities planning to conduct small UAS operations. Strict adherence to all of the operations and procedures outlined in this document (as well as those that are evolving) must be followed, along with compliance with all federal, state, and local ordinances. A safety perimeter should also always be established around the flight operations area.

SECTION 6

Small UAS Program Structure

A structured small UAS program is mandatory to support safe and efficient flight operations and ensure compliance with local, state, and federal laws, rules, statutes and regulations.

Proper organizational structure within a small UAS program consists of defined tasks for managers, crew and support personnel. It also allows subordinate personnel to temporarily "step up" into higher roles, as needed, in order to ensure the sustainability of the small UAS program. The recommended structured approach for designated personnel duties is the following:

- Program Manager (PM)
- Operations Manager (OM)
- Remote Pilot in Command (RPIC)
- Payload Operator (PO)
- Visual Observer (VO)
- Maintenance Technician (MT)

The small UAS program will require a policy and procedure for the charging and the storage of multiple batteries for the vehicles being operated. Safety procedures may be required for the bulk storage of batteries in order to comply with local, state and federal statutes for the handling of hazardous materials.

Federal regulations stipulate that employers provide their employees with Personal Protective Equipment (PPE) to protect them against contact with an identified hazards in the workplace — in this case the aircraft during flight operations. This equipment includes:

- Eye and Face Protection
- Head Protection

- Foot Protection
- Hand Protection

A PPE training program must ensure that employees are familiar with, and properly fitted for the protective equipment which must be properly maintained or replaced as necessary.

Identifying and monitoring data points within a small UAS Program will provide a trend analysis for the program's managers and support program justification decisions such as budgeting and planning. Identifying positive and negative trends allows timely and justified program direction changes when deemed necessary.

Small UAS Program Structure

The small UAS program consists of defined tasks for managers, crew, and support personnel. Proper organizational structure allows subordinate personnel to temporarily step up into higher roles, as needed, in order to ensure the sustainability of the small UAS program. A successful small UAS program designates personnel titles and duties according to FAA requirements. The following is a recommended approach for designated personnel of a small UAS program.

Program Manager (PM)

The Program Manager (PM) represents the small UAS Program and is the direct liaison with the FAA. The PM is solely responsible for the program's administrative adherence to requirements, policies, regulations, and success. The following is a list of requirements and recommendations for a PM.

PM Responsibilities

- Required
    - Plans and submits monthly reports (Monthly COA Report)
        - Provides additional internal (program) reports as required
    - Submits information for COA (routine) and ECOA (urgent) requests
        - Request a COA or ECOA for new small UAS operations
    - Submits FAA-required initial and renewal documents for the program
        - Submits Aircraft Registration documents

- Ensures all small UAS operational planning requirements are met
- Files all required incident and accident reports as applicable
* Recommended
  - Coordinates and attends all small UAS meetings and planning sessions
    - Identifies the point of contact within the organization operating the small UAS. Facilitates the program's interaction within the organization's internal dynamics when coordinating and planning with outside agencies
  - Compiles organizational small UAS program data points
    - Analyzes and assesses the small UAS program and supporting decisions relating to budget and operational cost forecast
  - Ensures training exercises are scheduled and executed
    - Implements processes, procedures, and policies are tested to ensure efficiency and effectiveness
    - Conducts monthly and annual audits of small UAS personnel individual training records (ITR)
    - Documents currency training, medical status, pilot overall proficiency and history
  - Conducts Semi-Annual Assessment of small UAS program
    - Identifies program benefits
  - Conducts a 3-year Benefit Trend Comparison Assessment
    - Identifies program benefits

Operations Manager (OM)

The Operations Manager (OM) ensures that all current and future small UAS operations are scheduled, planned, and supported. The OM also ensures that NOTAMs and post-operational reports are compiled and submitted in a timely manner.

Responsibilities

* Required
  - Ensures all small UAS operational planning requirements are met
    - Complies with all FAA and ATC coordination and communication requirements
  - Ensures currency of designated small UAS personnel

- Ensures currency training is scheduled and executed in accordance with the PM's intent.
- Obtains all required permissions and permits from territorial, state, county or city jurisdictions. This includes local law enforcement, fire, or other governmental agencies when applicable

- Recommended
  - Reports to the PM
    - Under the direction of the PM, receives authority from the PM as the supervisor of the small UAS program
  - Assigns crew to support small UAS operations
    - Provides a single point of contact (POC) for all crew, equipment and task assignments
  - Ensures all flights are in compliance with all regulations and laws
    - Provides a single POC for oversight of the remote pilot in command (RPIC). Must comply with local and state laws
  - Assists the PM duties and responsibilities as needed
  - Assists with small UAS operational coordination and meetings
    - Assists with submission of the monthly COA report
      - Additional confidential reports may be required
  - Assists with the compilation of small UAS Program data points
    - Data points are useful for the overall analysis and assessment of the small UAS program. They support decisions related to budgeting and operational cost forecast
  - Recommends personnel for small UAS designations and training
    - Recommendations of personnel skill development ensure the health of the small UAS program
  - Assists the PM during monthly and annual audits of small UAS personnel ITRs
    - Ensures that training and medical status requirements are met
  - Assists the PM in filing incident and accident reports
  - Assists the PM with the semi-annual assessment of the small UAS program
    - Facilitates program benefit trend identification
    - Assists the PM during a 3-year benefit trend comparison assessment
    - Facilitates program benefit trend identification

Remote Pilot in Command (RPIC)

The remote pilot in command (RPIC) is the primary crewmember and the sole authority overseeing aircraft operations pertaining to all phases of the small UAS flight.

RPIC Responsibilities

The RPIC is responsible for, but not limited to, the following:

- Required
    - The Code of Federal Regulations, Title 14, Part 107 (14 CFR, Part 107) requires that the flight controls of a small UAS be manipulated by a person with a remote pilot in command certificate, or by a person under the direct supervision of a licensed remote pilot in command who has the ability to immediately take the flight controls
    - The remote pilot in command must have passed either an initial or recurrent aeronautical written knowledge test within the previous 24 calendar months
    - Persons holding a pilot certificate (other than a student pilot certificate) for manned aircraft issued under CFR 14 Part 61 and meet the biennial flight review requirements must have completed an initial or recurrent online training course within the past 24 calendar months to obtain a RPIC rating
    - Prior to, or during any flight operation of a small UAS, a remote pilot in command must be specifically designated
    - The remote pilot in command is directly responsible for and is the final authority for the small UAS operation
    - The remote pilot in command must ensure that the small UAS will not pose a hazard to people, aircraft, or property in the event of a loss of control of the aircraft
    - The remote pilot in command must ensure that the small UAS operation complies with all regulations, flight rules and limitations
    - The remote pilot in command must have the ability to maintain control of the small UAS to ensure compliance with regulations, flight rules and limitations
    - During an in-flight emergency, the remote pilot in command may deviate from any rule to the extent necessary to meet that emergency
    - The RPIC provides supervision for the crew assigned to the mission

Each remote pilot in command who deviates from a rule must document the deviation and the reason for the deviation. That documentation may be used to complete and file a report upon the request of the FAA.

Any person who knows or has reason to know that he or she has a physical or mental condition that would interfere with the safe operation of a small UAS may not act as a remote pilot in command, visual observer, or directly participate in its operation.

- Recommended
    - Reports to the OM and accepts assignments
    - Ensures the safety of all small UAS crew and equipment
    - Reviews the quality of aerial data collected
    - Supplies data reports to the point of contact
        - Forwards all relevant data to PM for organization and reporting.
    - Assists PM with all required program renewals

Payload Operator (PO)

Payload Operators (PO) operate the small UAS payload during flight. The PO may also help maintain situational awareness, safety of flight and ground operations.

PO Responsibilities

The PO is responsible for, but not limited to, the following:

- Required
    - Collects aerial data for the crew
    - Assists the RPIC and maintains communications during all aspects of small UAS operations
    - Assists the RPIC during all required small UAS reporting
- Recommended
    - Reports to the OM when not assigned to a crew. When assigned to a crew, reports directly to the RPIC
    - Ensures the quality of aerial data collected during the small UAS operation and post processing
        - Assists in the analysis of the aerial data collected to ensure proper task completion
    - Supplies all required data points to the RPIC

Visual Observer (VO)

The primary duty of a Visual Observer (VO) is to help the RPIC maintain visual line of sight (VLOS) with the aircraft during the operation. The VO must maintain communications with the RPIC at all times.

VO Responsibilities

The VO is responsible for, but not limited to, the following:

- Required
    - Reports to the OM when not assigned to a crew; when assigned to a crew, reports directly to the RPIC
        - Assists the RPIC and maintain visual line of sight of the small UAS
        - Maintains continual communications with the RPIC during small UAS operations
- Recommended
    - Assists the RPIC with supplying all required data points

Maintenance Technician (MT)

A Maintenance Technician (MT) may be required to maintain one or a fleet of small UAS. In some small UAS Programs, the duties of a Maintenance Technician may be fulfilled by the RPIC, VO, or PO depending on their skills, experience, and qualifications.

MT Responsibilities

The MT is responsible for, but not limited to, the following:

- Required
    - Performing maintenance, inspections, and repairs on small UAS
    - Keep the maintenance logbooks complete and current
- Recommended
    - Reports to the OM when not assigned to a crew. When assigned to a crew, reports directly to the RPIC
    - Maintains the small UAS in an airworthy condition
    - Assist the RPIC with the functional flight test documentation

Storage and Accountability

Lithium Battery Storage

Special attention must be taken when storing lithium-ion and lithium polymer batteries. A hazardous material (HAZMAT) program should be created to address the proper storage and handling of lithium batteries. It is imperative that this storage be done in accordance with the manufacturer's recommendations and in compliance with all statutory regulations. Damaged batteries should be disposed of immediately, in compliance with local ordinances and only at authorized disposal facilities. Lithium batteries must be handled properly in order to ensure the safety of personnel, equipment, and facilities.

FAA regulations restrict the transportation of lithium batteries on commercial airliners. Spare lithium batteries must be carried in carry-on baggage only. Lithium batteries are limited to a rating of 100 watt hours (Wh) per battery. With airline approval, passengers may also carry up to two spare larger lithium batteries (101-160 watt hours). Batteries must be protected from damage and battery terminals must be covered and protected from short circuits.

Accountability

The small UAS program should maintain an updated small UAS "status" whiteboard to assist planners and schedulers with accountability, planning, scheduling, and asset/crew responsibilities. The small UAS whiteboard should contain:

- Small UAS identification number (N-number)
- Operational status
- Location
- Crew assigned
- Date/time updated

Small UAS Personal Protective Equipment (PPE)

Personal protective equipment (PPE) is important to the safety of the crew as well as to the identification of personnel during small UAS operations. Reflective safety vests, eye protection and first aid supplies must be made available and utilized during small UAS operations. All non-participating personnel must be protected with appropriate barriers.

The United States Department of Labor, Occupational Safety and Health Administration (OSHA) requires employers to provide safe work areas and reduce the likelihood of illness and injury to their employees from hazards in the workplace. OSHA standards are published in Title 29 of the Code of Federal Regulations. Also included in 29 CFR are the standards for command and control – the Incident Command System, the role of the Incident Commander and Safety Officer, and the development of a Site Safety Plan.

The potential hazards to crew in small UAS operations may include:

- Being struck by the aircraft during flight; specifically takeoff and landing
- Making contact with moving rotor(s)
- Electrical shock from battery chargers
- Burns from overheated electrical components

Work practice controls include establishing policies and procedures that reduce the employees' potential exposure to hazards. A work practice control also includes establishing and maintaining a PPE program that includes providing PPE, training in proper donning and doffing of the PPE, and proper maintenance and replacement of the PPE.

When working near small UAS operations, the following work place controls should be considered:

- The necessity of the mission and the assurance that flight parameters can be maintained.
- An exclusion zone must be established. Only crew with specific responsibilities for small UAS operations should be allowed access. Under the Incident Command System (ICS), the establishing of Operations Areas is the responsibility of the Incident Commander, based on the recommendations of the Operations Section Chief and Safety Officer. The Incident Commander should designate a qualified Safety Officer to develop a Site Safety Plan. The Safety Officer determines the need for PPE and advises the Incident Commander on the type of PPE required for the safety of personnel. Mission-specific PPE may be required for certain small UAS operations (e.g., personal flotation devices when working around water).
- The Safety Officer also monitors ground operations that include charging and replacing batteries according to manufacturer guidelines.

American National Standards Institute (ANSI) approved PPE must be provided to protect employees from contact with the aircraft during operations:

- Impact helmets with face shields
- Impact resistant eye shields that allow for wearing eyeglasses and/or contact lenses
- Cut-resistant gloves for handling aircraft when rotors are activated
- Sawyers-type leather apron for those who will come in direct contact with the aircraft while it is in operation

The applicable ANSI standards are:

- Eye and Face Protection: ANSI Z87.1-1989 (USA Standard for Occupational and Educational Eye and Face Protection).
- Head Protection: ANSI Z89.1-1986.
- Foot Protection: ANSI Z41.1-1991.

The PPE training program must ensure that employees are familiar with, and properly fitted for the PPE, and that the PPE is properly maintained. A PPE inspection program must be established and procedures written for replacing PPE when deemed necessary.

Small UAS Reporting and Data Points

Data points within a small UAS program provide a trend analysis for the program managers and support program justification decisions such as budgeting and planning. It is important to manage data efficiently within a program in order to analyze program benefits. Identifying positive and negative trends allows timely and justified program direction changes when deemed necessary.

SECTION 7

Small UAS Flight Operations

A Flight Operations and Procedures Manual (FOPM) must be developed to outline proper procedures when operating a small UAS in the National Airspace System (NAS). All flight crew and ground personnel must follow the procedures outlined in the FOPM.

Mission Planning

Mission planning must include normal operations, and include emergency operating procedures in order to mitigate unforeseen issues and situations. The following must be considered:

- Defining specifications and limitations of small UAS
- Studying map of the mission area
- Determining the type of imagery and data to be collected
- Site Safety Plan and perimeter control
- Selecting flight crew and ground support personnel
- Estimating costs and expenses for the mission

Flight Rules

Under Title 14 CFR Part 107, the person operating the small UAS must comply with the following:

- Small UAS must not exceed 87 knots in groundspeed
- Small UAS must not exceed altitude of 400 feet
- Flight visibility: within 3 statute miles
- Maintain a minimum distance from clouds: 500 feet below and 2,000 feet horizontally

- Small UAS must yield right of way to all aircraft
- Must avoid collision hazard
- May not operate over non-participating personnel

The Remote Pilot in Command must conduct a preflight inspection prior to each flight to determine if the small UAS is in a condition for safe operation.

The Remote Pilot in Command will terminate the flight and land immediately if communication is lost or if any unauthorized personnel or vehicles enter the operations area.

Flight Operations and Procedures Manual

A Flight Operations and Procedures Manual (FOPM) must be developed. The procedures in the FOPM will ensure compliance with regulations established by the FAA for conducting small UAS flight operations within the National Airspace System (NAS).

The FOPM outlines procedures that the flight crew and support personnel must follow. The contents of the manual are written to ensure a comprehensive understanding of the duties and responsibilities required for safe flight operations.

The information in the FOPM is based on data available at the time of publication and must be supplemented and kept current. All flight crew and ground personnel must follow all procedures outlined in the FOPM.

Mission Planning

Mission planning for each small UAS operation is dictated by a large number of variables and determined by the capabilities and limitations of the small UAS, crew, personnel and terrain. Operators must conduct detailed mission planning that includes a detailed assessment of the small UAS, crew, personnel and terrain. Mission planning must include normal operating procedures as well as emergency operating procedures in order to mitigate unforeseen issues and situations. The following criteria must be considered:

- Defining specifications and limitations of small UAS
- Studying a map of the planned mission area
- Determining the type of aerial imagery and data to be collected
- Site Safety Plan and perimeter control
- Selecting flight crew and ground personnel

- Estimating costs and expenses for the small UAS mission
- Coordinating with Air Traffic Control (ATC) or airport authority as necessary

Flight Rules

Title 14 of the Federal Code of Regulations (14 CFR), Part 107 requires that a Remote Pilot in Command and/or the person manipulating the flight controls of the small UAS comply with all of the following limitations when operating a small UAS:

1. The groundspeed of the small UAS may not exceed 87 knots (100 miles per hour).
2. The altitude of the small UAS cannot be higher than 400 feet above ground level unless the small UAS is flown within a 400-foot radius of a structure, and does not fly higher than 400 feet above the structure's immediate uppermost limit.
3. The flight visibility, as observed from the location of the control station, must be less than 3 statute miles. Flight visibility is the average slant distance from the control station at which prominent unlighted objects may be seen and identified by day, and prominent lighted objects may be seen and identified by night.
4. The minimum distance of the small UAS from clouds must be no less than 500 feet below the cloud and 2,000 feet horizontally from the cloud.
5. Each small UAS must yield the right of way to all aircraft, airborne vehicles, and launch and reentry vehicles. Yielding the right of way means that the small UAS must give way to the aircraft or vehicle, and may not pass over, under, or ahead of it unless well clear.
6. No person may operate a small UAS so close to another unmanned aircraft so as to create a collision hazard.
7. No person may operate a small UAS over a human being unless that human being is directly participating in the flight operation or located under a covered structure (or inside a stationary vehicle) that can provide reasonable protection from a falling small UAS.

Prohibited Operations

1. No person may act as a Remote Pilot in Command or Visual Observer for more than one small UAS at a time.
2. Small UAS shall not be operated in Class B, C, D or E airspace unless prior authorization from Air Traffic Control (ATC) has been issued with a certificate of waiver.

3. Small UAS aircraft shall not be operated in a manner that interferes with operations and traffic patterns at any airport, heliport, or seaplane base.

4. Small UAS shall not be operated in prohibited or restricted areas unless permission has been granted from the controlling agency.

5. Small UAS may not carry hazardous material. For purposes of this section, the term hazardous material is defined in Title 49 CFR Part 171.8.

6. Small UAS shall not be operated in a careless or reckless manner so as to endanger the life or property of another.

7. The operator of a small UAS shall not allow an object to be dropped during flight in a manner that creates an undue hazard to persons or property.

8. External load operations are only allowed if the object being carried by the small UAS is securely attached and does not adversely affect the flight characteristics or controllability of the aircraft.

9. Small UAS flight operations are not permitted from a moving aircraft.

10. Flight operations from a moving land or water-borne vehicle are only permitted if the small UAS is flown over a sparsely populated area and is not transporting another person's property for compensation or hire.

11. A person acting as a Remote Pilot in Command must comply with all temporary flight restrictions in areas designated by any Notice to Airmen (NOTAM).

Preflight Inspection

The Remote Pilot in Command must conduct a preflight inspection prior to each flight to determine if the small UAS is in a condition for safe operation.

The Remote Pilot in Command must ensure that any object attached or carried by the small unmanned aircraft is secure and does not adversely affect the flight characteristics or controllability of the aircraft.

The Remote Pilot in Command must ensure that there is enough available power for the small UAS to operate for the intended operational time.

If the preflight inspection reveals that the small UAS is not in a condition for safe operation or that the control link is not functioning properly, the Remote Pilot in Command shall not commence flight operations until the small UAS is in a condition for safe operation and any and all control link deficiencies have been corrected. The Remote Pilot in Command will document any deficiencies that prohibit operations that are found during the Preflight Inspection.

Operational Assessment

Prior to conducting any flight operations, the Remote Pilot in Command must conduct an operational assessment. The assessment must include the following:

1. Local weather conditions
2. Local airspace and flight restrictions
3. The location of persons and property
4. Ground hazards

Briefings and Security

The Remote Pilot in Command must ensure that all persons directly participating in the small UAS are informed about the operating conditions, emergency procedures, contingency procedures, roles and responsibilities, and potential hazards.

The briefing and safety meeting shall be conducted for all persons directly participating in the flight operations, including emergency, safety and security personnel and shall cover:

- Perimeter of flight operations area
- Aircraft launch & landing zones
- Radio communications
- Takeoff procedures
- Approach and landing procedures
- Flight Termination procedures
- Emergency procedures
- Risks to participating personnel
- PPE that is required for participating personnel
- Procedures for mitigating risk to participating personnel
- How to control nonparticipating persons
- Local governmental limitations or restrictions, if any

Prior to any flight operations, a plan for securing the flight operations area from all unauthorized persons and vehicles will be developed. Provisions will be made to immediately terminate flight operations if unauthorized individuals or vehicles enter the area.

Security and ground personnel will keep directly participating personnel continuously apprised of the status of the operations utilizing radio communication, oral, visual, or a combination thereof.

Flight Termination

The Remote Pilot in Command will terminate the flight operation and land immediately if any of the following situations occur:

1. Non-participating personnel enter the flight operations area.

2. Participating personnel violate their predetermined distance and location from the small UAS flight path.

3. Unauthorized vehicles enter the flight operations area.

4. The small UAS inadvertently breeches the lateral or vertical boundary of the flight operations area.

5. Communication with the Visual Observer is lost.

In the event the GPS signal is lost during flight operations, the Remote Pilot in Command shall return the small UAS to a pre-determined location, either manually or autonomously, and land it, or recover the small UAS in accordance with the emergency procedures outlined in the Aircraft Flight Manual.

SECTION 8

Small UAS Post Flight Operations

In order to comply with FAA regulations, a post mission report providing a brief summary of the small UAS flight activity must be generated. This summary must include all the flight documentation collected during the mission, including any abnormal flight behavior, emergency procedures and record of compliance.

FAA Regulations define the parameters of accidents and incidents and mandate that these accidents or incidents must be reported no later than 10 days after they occur.

The post mission report should include any actions taken to address abnormal occurrences during a small UAS flight mission.

The Remote Pilot in Command shall document the equipment and crew involved in small UAS missions in a logbook and designate the small UAS as fully mission capable (FMC) or non-mission capable (NMC).

A small UAS status tracking board can provide a quick reference to equipment status and location, mission and flight crew assignments.

Tracking the cost of each small UAS flight operation with a Budget Document helps determine any future small UAS acquisitions and assists with documentation for potential FEMA reimbursements for disaster response.

Post Flight Mission Report

A post flight mission report must be generated to provide a brief summary of the small UAS flight activity. This summary must include the following:

- Small UAS registration number
- Small UAS serial numbers
- Weather conditions
- Operating altitudes
- Summary of aerial data collected
- Desired data points
- Abnormal flight behavior
- Emergency procedures
- Record of legal compliance — NOTAM number and information published, permissions obtained from property owners to overfly their property, etc.
- ICS #220 – Air Operations Summary for missions conducted during emergency response. (For training and experience with the documentation, it would be helpful to complete the Air Operations Summary even for non-emergency missions.)

Accident and Incident Documentation

A Remote Pilot in Command must report within 10 days to the FAA any small UAS operation involving the following:

1. Serious injury to any person or any loss of consciousness.
2. Damage to any property, other than the small unmanned aircraft, unless the cost of repair does not exceed $500; or the fair market value of the property does not exceed $500 in the event of total loss.

Flight Logbook

The RPIC should document all pertinent information about the small UAS, flight conditions, type of mission and mission parameters in a flight logbook. Flight logbooks should include:

- Registration number
- Type and location of small UAS damage, if any
- Type of replacement parts needed, if any
- Number of flights
- Duration of flights
- Assigned crew

The small UAS equipment status can be designated as fully mission capable (FMC) or non-mission capable (NMC). This information is useful in planning future small UAS program activities, budgeting, identifying small UAS life cycle and maintenance data.

Small UAS Status Tracking Board

The small UAS status tracking board is a quick reference pertaining to location, operational status, and mission assignment. Updated information is vital as the status board is viewed by small UAS program planners, managers and crew. The information included on a small UAS status board must include the following:

- Small UAS registration equipment number
- Small UAS operational status (FMC, NMC)
- Location of equipment
- Mission assignment and dates
- Assigned crew

Small UAS Budget Document

The small UAS Budget Document tracks the original equipment cost as well as the cost associated with each flight. Tracking the cost per flight supports decisions regarding the acquisition of additional assets or the deactivation of existing small UAS. The information gathered must meet all reporting requirements, as well as FEMA reimbursement guidelines.

SECTION 9

Small UAS Training

All flight crew of a small UAS Program emergency response team should be trained in the National Incident Management System (NIMS) and the Incident Command System (ICS). In addition to proper FAA certification, the flight crew must demonstrate proficiency and safe operation of small UAS prior to conducting operations.

Public and private sector training for small UAS flight crew, must include the following NIMS/ICS training courses to be considered qualified for emergency response:

IS-100.B: Introduction to the Incident Command System

IS-200.B: ICS for Single Resources and Initial Action Incidents

IS-700.A: Introduction to the National Incident Management System (NIMS)

IS- 800.B: Introduction to the National Response Framework

Training manuals should be created to assess and confirm that the flight crew has attained and maintains the knowledge and proficiency required for safe small UAS operations.

The FAA requires Remote Pilot in Command (RPIC) personnel to undergo initial and recurrent training in aeronautical knowledge and small UAS training.

Visual Observers (VO) must be proficient at weather briefing, hazard identification and communications procedures. A Visual Observer who is also qualified as RPIC would add safety benefits to flight operations. A qualified VO could assume the controls, operate and land the unmanned aircraft in the event the primary RPIC becomes incapacitated.

The Payload Operator (PO) must be proficient at the specific procedures required for operating the sensors onboard the small UAS. The PO should be able to assume the role of qualified Visual Observer in the event the designated VO is required to assume the role of RPIC.

NIMS/ICS Training

At a minimum, Visual Observers and other members of the flight crew must receive the same level of training as NIMS recommends for a "Field Observer." If pilots and other supervisory personnel are established, they should attain the same training levels as their counterparts — specifically as Single Resource Unit Leaders at a minimum. As currently required for all emergency responders, both public and private sector training for small UAS flight crew must include at least the following NIMS/ICS training courses:

IS-100.B: Introduction to the Incident Command System

This course introduces the Incident Command System (ICS) and provides the foundation for higher level ICS training. This course describes the history, features and principles, and organizational structure of the Incident Command System. It also explains the relationship between ICS and the National Incident Management System (NIMS).

IS-200.B: ICS for Single Resources and Initial Action Incidents

This course is designed to enable personnel to operate efficiently during an incident or event within the Incident Command System (ICS). ICS-200 provides training on and resources for personnel who are likely to assume a supervisory position within the ICS.

IS-700.A: National Incident Management System (NIMS), An Introduction

This course introduces and overviews the National Incident Management System (NIMS). NIMS provides a consistent nationwide template to enable all government, private-sector, and nongovernmental organizations to work together during domestic incidents.

IS- 800.B: National Response Framework, An Introduction

This course introduces participants to the concepts and principles for the National Response Framework. This course is intended for government executives, private-sector and non-governmental organizations (NGO) leaders and emergency management practitioners.

RPIC Training

Training manuals should be created to confirm the flight crew has attained and maintains the knowledge and proficiency for the safe operation of the small UAS to be operated. Training manuals and evaluations should include the skills to be demonstrated in a practical examination. A flight test conducted by a qualified third party should be administered to confirm the RPIC meets a required level of flight proficiency.

The FAA has established two separate knowledge requirements required for the RPIC based on whether the applicant already has a pilot certificate:

- In order to be eligible for an FAA remote pilot certificate with a small UAS rating under 14 CFR, Part 107, a person must successfully complete an initial aeronautical knowledge test covering the areas of knowledge further specified below.

- If a person holds a pilot certificate (other than a student pilot certificate) issued under 14 CFR part 61, and meets the biennial flight review requirements, to obtain an RPIC rating that person must successfully complete an initial training course covering the areas of knowledge further specified below.

INITIAL AND RECURRENT AERONAUTICAL KNOWLEDGE TESTS

The initial aeronautical knowledge test covers the following subjects:

1. Applicable regulations relating to small unmanned aircraft system rating privileges, limitations, and flight operations
2. Airspace classification, operating requirements, and flight restrictions affecting small unmanned aircraft operations
3. Aviation weather sources and effects of weather on small unmanned aircraft performance
4. Small unmanned aircraft loading
5. Emergency procedures
6. Crew resource management
7. Radio communication procedures
8. Determining the performance of small unmanned aircraft
9. Physiological effects of drugs and alcohol
10. Aeronautical decision-making and judgment

11. Airport operations
12. Maintenance and preflight inspection procedures

A recurrent aeronautical knowledge test covers the following areas of knowledge:

1. Applicable regulations relating to small unmanned aircraft system rating privileges, limitations, and flight operations
2. Airspace classification and operating requirements and flight restrictions affecting small unmanned aircraft operations
3. Emergency procedures
4. Crew resource management
5. Aeronautical decision-making and judgment
6. Airport operations
7. Maintenance and preflight inspection procedures

INITIAL AND RECURRENT TRAINING COURSES

An initial training course covers the following areas of knowledge:

1. Applicable regulations relating to small unmanned aircraft system rating privileges, limitations, and flight operations
2. Effects of weather on small unmanned aircraft performance
3. Small unmanned aircraft loading
4. Emergency procedures
5. Crew resource management
6. Determining the performance of small unmanned aircraft
7. Maintenance and preflight inspection procedures

A recurrent training course covers the following areas of knowledge:

1. Applicable regulations relating to small unmanned aircraft system rating privileges, limitations, and flight operations
2. Emergency procedures
3. Crew resource management
4. Maintenance and preflight inspection procedures

VO Requirements

The Visual Observer (VO) training manual must outline the process and experience that the VO must have in order to be confirmed for this position. Besides defining the areas of knowledge required, the VO training manual serves as a reference for the experienced VO.

The VO must be present for all flight operations including training, proficiency, currency, and post-maintenance. The VO must be proficient at weather briefing, hazard identification and communications procedures.

The VO should take a written test documenting that all the VO qualification criteria have been met. The VO written test must be implemented by the small UAS OM.

The VO can offer safety advantages if the VO is trained and qualified as a RPIC. The VO could assume the controls, operate and land the unmanned aircraft in the event the primary RPIC becomes incapacitated or is unable to continue operations. Under these conditions, the continuation of the mission should be evaluated. Termination of the mission, as described elsewhere is strongly advised. Should the VO be required to assume the controls of the unmanned aircraft, the Payload Operator should assume the function of Visual Observer until the mission is terminated.

Payload Operator (PO) Requirements

The Payload Operator (PO) training manual contains the specific procedures required for operating the sensors onboard the small UAS. The PO training manual must outline the process and experience required for this position.

The PO must take a written test documenting that all qualification criteria have been met prior to being confirmed as a member of the flight crew. The PO must also be knowledgeable about safety and communications procedures.

SECTION 10

Small UAS Maintenance

Proper maintenance of a small UAS is essential to safe and efficient flight operations. Trained personnel must be designated to conduct periodic inspections and required maintenance.

All manufacturer scheduled maintenance instructions should be followed in the interest of achieving the longest and safest service life of the small UAS. If scheduled maintenance procedures are not provided by the manufacturer, procedures for documenting repair, modification, overhaul or replacement of a system component resulting from normal flight operations should be developed.

Maintenance on small UAS must be performed and logged on a regular basis. When selecting the person in charge of maintenance performance and record-keeping, the Program Manager must decide which of the following best meets its requirements:

- RPIC
- Maintenance Technician
- Manufacturer or designated repair facility

Electric, small unmanned aircraft have limited flight endurance and short flight times. It is recommended that a complete aircraft inspection be performed every 100 cycles (a takeoff and landing constitutes one cycle) in addition to the required annual inspection.

Record-keeping that includes a record of all periodic inspections, maintenance, preventative maintenance, repairs and alterations performed on the small UAS should be retrievable from either hardcopy and/or electronic logbook format. Proper record-keeping of small UAS maintenance reinforces responsibilities for airworthiness through systematic condition for safe flight determinations.

After necessary maintenance has been performed and the small UAS is found to be in safe condition for flight, the Remote Pilot in Command should conduct a functional flight test and document the test in the maintenance logbook section of the aircraft flight manual before it is returned to service.

Scheduled Maintenance

All small UAS must be maintained in a condition for safe operation. Some small UAS manufacturers may provide recommendations for scheduled maintenance of the aircraft and associated system equipment. There may be components of the small UAS that are identified by the manufacturer to undergo scheduled periodic maintenance or replacement, based on time-in-service limits (such as flight hours, cycles, and/or calendar days). All manufacturer scheduled maintenance instructions should be followed in the interest of achieving the longest and safest service life for the small UAS.

When scheduled maintenance procedures have not been recommended or provided by the small UAS manufacturer, maintenance procedures must be developed to document any repair, modification, overhaul, or replacement of a system component resulting from normal flight operations. Also, the time-in-service for that component at the time of the maintenance procedure should be recorded. When available, a small unmanned aircraft system's component manufacturer's life limit requirements should be followed. Over time, a reliable maintenance schedule for the small UAS and its components can be developed.

Maintenance Personnel

Regular maintenance not only improves safety of flight by preventing accidents, but also assists in detecting recurring problems, ultimately reducing operational liability. Maintenance records document the work performed, the technician assigned and the date completed. Maintenance on small UAS must be performed and logged on a regular basis. When selecting the person in charge of maintenance performance and record-keeping, the Program Manager must decide which of the following best meets its requirements:

- RPIC
- Maintenance Technician
- Manufacturer or designated repair facility

Remote Pilot in Command (RPIC) — Maintenance of the small UAS may be performed by the RPIC if that person meets the required qualifications and has received the necessary training. Performing preflight inspections and operating the small UAS makes the RPIC the most familiar with the aircraft. Since discrepancies may be discovered during the preflight inspection, the RPIC must have the ability to perform minor maintenance on the aircraft. The ability to conduct minor repairs in the field is a qualification that must be considered when selecting an RPIC. If the RPIC is not the designated maintenance technician, any repairs performed in the field should be documented and checked by the maintenance technician after the flight.

Maintenance Technician – Operating a fleet of small UAS may require a dedicated small UAS technician on staff. Record-keeping requirements for performed maintenance on a small UAS fleet can be substantial.

Without certification requirements for a small UAS technician, a person's qualifications must be evaluated for this critical position. Experience and training with the small UAS assets may guide the selection of personnel.

Manufacturer – The small UAS manufacturer may be the best resource for extensive overhaul or repair procedures. This option places the operator at a disadvantage because of the time required to ship the small UAS to the manufacturer and the time required for their technicians to perform the necessary repairs.

Technical support may or may not be offered by a manufacturer. This is another matter for consideration prior to equipment acquisition.

Inspection Requirements

As required by Federal Aviation Regulations, all civil aircraft of U.S. registry must undergo a complete inspection (annual) each twelve calendar months. In addition to the required annual inspection, aircraft operated commercially (for hire) must have a complete aircraft inspection every 100 hours.

Given the fact that electric, small unmanned aircraft have limited flight endurance and short flight times, it is recommended that a complete aircraft inspection be performed every 100 cycles (a takeoff and landing constitutes one cycle) in addition to the required annual inspection.

This increased inspection schedule will ensure the highest level of safety and reliability while conducting flight operations with a small UAS.

Record-keeping Requirements

Record-keeping that includes a record of all periodic inspections, maintenance, preventative maintenance, repairs, and alterations performed on the small UAS could be retrievable from either hardcopy and/or electronic logbook format. Record-keeping of documented maintenance and inspection events reinforces responsibilities for airworthiness through systematic condition for safe flight determinations. Maintenance and inspection record-keeping provides retrievable empirical evidence of vital safety assessment data defining the condition of safety-critical systems and components supporting the decision to conduct flight operations. Record-keeping of a small UAS may provide essential safety support for commercial operators that may experience rapidly accumulated flight operational hours/cycles. Methodical maintenance and inspection data collection can prove to be very helpful in the tracking of small UAS component service life and structural failure events.

Any unmanned aircraft that has undergone maintenance, alterations, or repair that affect its operation or flight characteristics must undergo a functional test flight in accordance with the operator's manual. The Remote Pilot in Command who conducts the functional test flight must make an entry in the maintenance log section of the Aircraft Flight Manual.

If during a preflight inspection the Remote Pilot in Command identifies a condition that affects the safe operation of the unmanned aircraft, the discrepancy found by the Remote Pilot in Command must be documented in the aircraft discrepancy log and the aircraft prohibited from operating.

After the necessary maintenance has been performed and the unmanned aircraft is found to be in safe condition for flight, the Remote Pilot in Command will conduct a functional flight test in accordance with the procedures outlined above. The Remote Pilot in Command must document the functional flight test in the maintenance logbook section of the aircraft flight manual before the aircraft is returned to service.

Conclusion

Developing, implementing, and managing a small UAS program involves serious considerations that must be addressed to meet both organizational and operational requirements. Properly trained personnel are essential to establishing a successful small UAS program. The lack of standardized flight training presents challenges for any organization or agency. Properly maintained small UAS are essential to safe and reliable operations, yet the FAA does not provide any specific guidelines or requirements for their maintenance. The operator and remote pilot in command must determine if the small UAS is in a safe condition for flight. Record-keeping and documentation for personnel and the small UAS are critical for the safety of any aviation program.

Federal law states that small UAS are aircraft and that compliance with all FAA regulations is mandatory for all flight operations in the National Airspace System. FAA regulations governing the operation of small UAS continue to evolve as this new technology becomes further integrated in the NAS. State and local governments are also creating privacy and trespassing laws which may impact small UAS operations in specific areas.

For these reasons, any agency or organization considering developing a small UAS program is strongly advised to seek guidance and counsel from consultants and attorneys with aviation experience. Subcontracting services from reputable small UAS providers can offer alternatives to the creation of a small UAS program. Whether an organization develops its own small UAS program or subcontracts the service, its adaptation of this new technology can potentially transform the way it meets its mission requirements.

References

Advisory Circular AC 00-1.1A Public Aircraft Operations, *Federal Aviation Administration*

Advisory Circular AC 00-45, Aviation Weather Services., *Federal Aviation Administration*

Advisory Circular 21-12C Application for U.S. Airworthiness Certificate, *Federal Aviation Administration*

Advisory Circular AC 45-2E Identification and Registration, *Federal Aviation Administration*

Advisory Circular 107-2, Small Unmanned Aircraft Systems, *Federal Aviation Administration*

Aeronautical Information Manual (AIM). *Federal Aviation Administration*

Emergency Management Institute.

FAA Modernization and Reform Act of 2012 (P.L. 112-095), *United States Congress*

NPRM, Small Unmanned Aircraft Systems 14 CFR Part 107, *Federal Aviation Administration*

Notice 8900.1, Education, Compliance, and Enforcement of Unauthorized Unmanned Aircraft Systems Operators. *Federal Aviation Administration*, 04 August 2015

Notice 8900.357, Inspection and Maintenance Program Requirements for Airworthiness Certification of Unmanned Aircraft Systems Operating Under 55 Pounds, *Federal Aviation Administration*

Notice JO 7210.891, Unmanned Aircraft Operations in the National Airspace System (NAS), *Federal Aviation Administration*, 25 November 2015

Order 8900, Volume 16, Unmanned Aircraft Systems, *Federal Aviation Administration*

Pilot's Handbook of Aeronautical Knowledge, *Federal Aviation Administration*

Rupprecht, Jonathan Drone Law for Public Safety Decision Makers, February 2016

Title 14 CFR Chapter I, Subsection E. *Federal Aviation Administration*

Title 14 CFR Part 1, Definitions and Abbreviations.

Title 14 CFR Part 48, Registration and Marking Requirements for Small Unmanned Aircraft.

Title 14 CFR Part 71, Designation of Class A, B, C, D, and E Airspace Areas; ATS Routes.

Title 14 CFR Part 73, Special Use Airspace.

Title 14 CFR Part 91, General Operating and Flight Rules.

Title 14 CFR Part 93, Special Air Traffic Rules.

Title 14 CFR Part 101, Moored Balloons, Kites, Amateur Rockets and Unmanned Free Balloons.

Title 14 CFR Part 107, Small Unmanned Aircraft Systems.

Title 47 CFR Part 87, Aviation Services.

Voluntary Best Practices for UAS Privacy, Transparency, and Accountability, *National Telecommunications and Information Administration (NTIA)*

About the Authors

This book was written by three leading experts in the small UAS, emergency response and aviation law industries in conjunction with Cabezon Group Inc.

Gus Calderon

Gus Calderon is an FAA-licensed Commercial Pilot with over 22 years of aviation experience. He has a multi-engine and instrument rating and has over 3,000 flight hours in high-performance aircraft. Gus developed and operated an FAA-certified Part 135 air charter company based in Carlsbad, CA. He owns a Beechcraft Bonanza and conducts aerial photography missions.

Gus has been building and flying remote-controlled aircraft for decades. In 2006 he began installing autopilots and cameras in fixed-wing aircraft. In 2010 he started custom building multi-rotor platforms for aerial photography and cinematography. As an early-adopter of civilian unmanned aircraft, he has conducted thousands of flight operations and has over a decade field experience. He currently has an FAA Exemption, Remote Pilot in Command certification and has conducted small UAS operations for several commercial industries.

Because of his experience in commercial aviation and knowledge of multi-rotor systems, Gus was selected as a member of Lady Gaga's design and development team for her "Flying Dress" which was named the Volantis. He designed and built the electric power system and conducted all flight testing of the Volantis. On November 10, 2013, Gus successfully flew Lady Gaga on the Volantis at the ArtPop album release in Brooklyn, New York.

As an advocate for the beneficial use of small UAS, Gus has demonstrated how aeronautical knowledge can improve their safe and reliable operation. Using his experience as an air charter operator, he has developed flight operating manuals and

Kevin C. Rolfe

Kevin C. Rolfe is a public service professional with 32 years of service in county fire rescue and has extensive knowledge of Incident Management in the field and in the Emergency Operations Center (EOC).

Mr. Rolfe has worked and provided leadership in all facets of incident management in the field as well as overseeing emergency management programs on the federal, state and local level. Mr. Rolfe was a member of the Governor's Statewide Working Group for Domestic Security for the State of Florida, and a leading member of the Region 3 Domestic Security Task Force in Northeast Florida — serving as the co-lead of the Infrastructure Committee and Training Committee. Working as the co-lead, the Training Committee established minimum training requirements for response and emergency management personnel pre-dating the FEMA-established training guidelines. Additionally, Mr. Rolfe has authored or has been a contributing author to numerous publications related to emergency incident management, including contributing to the State of Florida Field Operations Guide; Prepare Florida for Terrorism Response and Instructor Guides; and the Florida Foundations of Emergency Management Academy.

Mr. Rolfe has provided Incident Command Systems (ICS) and Federal Emergency Management Agency (FEMA) training for multiple disciplines on the federal, state and local public sector, as well as for the private sector. Mr. Rolfe has conducted training programs and workshops at the Florida Governor's Hurricane Conference, Florida Emergency Preparedness Association, Florida Fire Chiefs' Association, and State of Florida Division of Emergency Management relating to the best practices in incident management and emergency management.

Jonathan Rupprecht

Drone Analyst, Newsweek, Politico, NPR, Marketwatch, The Independent, Motherboard, and many other sources have cited or quoted Jonathan B. Rupprecht who is a drone lawyer and a commercial pilot with single-engine, multi-engine, and instrument ratings. He is also an airplane flight instructor and instrument flight instructor. Jonathan obtained a B.S. from Embry-Riddle Aeronautical University, Magna Cum Laude, and a J.D. from Florida International University School of Law. He is currently an adjunct faculty professor for Embry-Riddle Aeronautical University teaching an Unmanned Systems Legal and Regulatory Compliance course. Jonathan authored a drone textbook being used at multiple universities around the U.S. He co-authored a legal treatise on unmanned aircraft that is being published by the American Bar Association. He is now currently working on the three Taylor v. FAA cases in the D.C. Circuit Court of Appeals challenging the FAA's drone registration regulations. Jonathan is currently practicing drone law in South Florida at his firm Rupprecht Law, P.A.

Maha Calderon - Editor

Maha Calderon received her Associates of Arts degree in Business Administration from McLennan Community College and her Bachelor of Arts degree in journalism from Texas State University. She worked for Harte Hanks News Service in Austin, Texas, covering the Texas State Legislature in its regular and special sessions, writing and editing news stories for seven Texas newspapers. Maha moved to California where she worked as a beat reporter covering San Juan Capistrano for the Daily Sun Post. Subsequently, she worked for the Gemological Institute of America (GIA) writing press releases, articles and newsletters for the jewelry industry. She was the primary commercial gemstone photographer for GIA and assisted with course development for the Graduate Gemologist program. Maha is an FAA Commercial Pilot and has worked as a Certified Flight Instructor. She is also a HAM Radio Technician

Cabezon Group

Based in Washington, D.C., Cabezon Group is an emergency management consulting firm that specializes in solving complex and mission critical challenges for the U.S. government. Cabezon Group is a leader in the field of emergency preparedness and response for federal agencies such as the Department of Homeland Security (DHS), Federal Emergency Management Agency (FEMA), and Department of Health and Human Services (HHS).

Cabezon Group focuses on small unmanned aircraft operations with an emphasis on integration, procedure and training protocols for a wide variety of emergency mission requirements. The company has working partnerships with leading authorities and organizations within the scientific and engineering communities. Cabezon Group delivers turnkey, customized solutions to the public and private sectors for seamless unmanned aircraft integration within mission and budget requirements.

Cabezon Group develops NIMS-compliant programs and provides training for various disaster and incident response scenarios, including guidance on FEMA reimbursement for federally declared incidents.

For more information, please visit www.cabezon.com.

www.ingramcontent.com/pod-product-compliance
Lightning Source LLC
Chambersburg PA
CBHW041103180526
45172CB00001B/92